LANDSCAPE AND AGENCY

Landscape and Agency explores how landscape, as an idea, a visual medium and a design practice, is organized, appropriated and framed in the transformation of places, from the local to the global. It highlights how the development of the idea of agency in landscape theory and practice can fundamentally change our engagement with future landscapes. Including a wide range of international contributions, each illustrated chapter investigates the many ways in which the relationship between the ideas and practices of landscape, and social and subjective formations and material processes, are invested with agency. They critically examine the role of landscape in processes of contemporary urban development, environmental debate and political agendas and explore how these relations can be analysed and rethought through a dialogue between theory and practice.

Ed Wall is the Academic Leader Landscape at the University of Greenwich, London, Visiting Professor at Politecnico di Milano (DiAP) and City of Vienna Visiting Professor 2017 (SKuOR) for urban culture, public space and the future – urban equity and the global agenda. Ed's research focuses on the design and theory of landscapes, public spaces and cities. He is the founding editor of the design research journal *Testing-Ground* (2015). In 2007 Ed established *Project Studio*. Award-winning projects have been published and exhibited widely, including at the Architecture Foundation, Royal Academy, Biennale of Landscape Urbanism, London Festival of Architecture and the Van Alen Institute.

Tim Waterman is senior lecturer and landscape architecture theory coordinator at the University of Greenwich, and a tutor at the Bartlett School of Architecture, UCL. He writes for a wide range of professional and academic publications on the subjects of power, democracy, taste, foodways and everyday life.

LANDSCAPE AND AGENCY

Critical Essays

Edited by Ed Wall and Tim Waterman

LONDON AND NEW YORK

First published 2018
by Routledge
2 Park Square, Milton Park, Abingdon, Oxon OX14 4RN

and by Routledge
711 Third Avenue, New York, NY 10017

Routledge is an imprint of the Taylor & Francis Group, an informa business

© 2018 Ed Wall and Tim Waterman

The right of the editors to be identified as the author of the editorial material, and of the authors for their individual chapters, has been asserted in accordance with sections 77 and 78 of the Copyright, Designs and Patents Act 1988.

All rights reserved. No part of this book may be reprinted or reproduced or utilised in any form or by any electronic, mechanical, or other means, now known or hereafter invented, including photocopying and recording, or in any information storage or retrieval system, without permission in writing from the publishers.

Trademark notice: Product or corporate names may be trademarks or registered trademarks, and are used only for identification and explanation without intent to infringe.

British Library Cataloguing-in-Publication Data
A catalogue record for this book is available from the British Library

Library of Congress Cataloging-in-Publication Data
A catalog record for this book has been requested.

ISBN: 978-1-138-12556-8 (hbk)
ISBN: 978-1-138-12557-5 (pbk)
ISBN: 978-1-315-64740-1 (ebk)

Typeset in Bembo
by Swales & Willis Ltd, Exeter, Devon, UK

Printed in the United Kingdom
by Henry Ling Limited

CONTENTS

List of figures	*vii*
Notes on contributors	*x*
Acknowledgements	*xiv*
Foreword	*xvi*
Murray Fraser	
Introduction: critical concerns of landscape	1
Ed Wall and Tim Waterman	
1 Landscapes of post-history	7
Ross Exo Adams	
2 Reciprocal landscapes: material portraits in New York City and elsewhere	18
Jane Hutton	
3 Agency, advocacy, vocabulary: three landscape projects	34
Jane Wolff	
4 The law is at fault? Landscape rights and 'agency' in international law	52
Amy Strecker	

5   How to live in a jungle: the (bio)politics of the park as
    urban model   65
    Maria Shéhérazade Giudici

6   Planetary aesthetics   78
    Peg Rawes

7   The closed landscapes of Sverdlovsk-44 and Krasnoyarsk-26   90
    Katya Larina

8   Rhythm, agency, scoring and the city   104
    Paul Cureton

9   Publicity and propriety: democracy and manners in Britain's
    public landscape   117
    Tim Waterman

10 The power of the incremental: agronomic investment in
    Lisbon's Chelas valley   131
    Jill Desimini

11 Post-landscape *or* the potential of other relations with the land 144
    Ed Wall

12 Activating equitable landscapes and critical design assemblages
    in Bangkok   164
    Camillo Boano and William Hunter

13 Agency and artifice in the environment of neoliberalism   177
    Douglas Spencer

    Afterword: landscape's agency   188
    Don Mitchell

Index   *193*

FIGURES

1.1	Stills from *Black Sea Files* by Ursula Biemann (2005)	15
1.2	*Now*, video installation by Chantal Akerman from a show of the same title of her work in Ambika P3, London (2015)	16
2.1	Sands Quarry in Vinalhaven, Maine, owned by the Bodwell Granite Co. Photograph published in *US Geologic Survey Bulletin* 313, 1907	20
2.2	South Gate House, Central Park, 1891 [neg #62683]	21
2.3	Granite cutters from the Bodwell Granite Co. marking the government contract (1872–1888) for the State, War, and Navy Building in Washington, DC	22
2.4	Houses and steel mills in Ambridge, Pennsylvania, 1938. Photographer: Arthur Rothstein (1915–1985)	23
2.5	Armed deputy sheriffs confronting picketers at the Spang, Chalfant Seamless Tube Company, Ambridge, Pennsylvania. Photograph published by World Wide Photos, 1933	24
2.6	Riverside Park at 82nd Street under construction, showing steel frames installed over the New York Central Railroad tracks, 1936	25
2.7	Logs and recently milled ipê (*Tabebuia* sp.) lumber, Belém, Brazil, 2012	27
2.8	Rainforests of New York demonstration on Phase 1 of the High Line, New York City, 2009	28
2.9	Reclaimed teak benches in Phase 2 of the High Line, New York City, 2011	29
3.1	Home page, *Gutter to Gulf* website	37
3.2	Water taxonomy: subsurface drainage system	38

3.3	A proposal for reimagining the urban water landscape	39
3.4	Sketch of vocabulary subjects, *Bay Lexicon*	41
3.5	A flash card's front and back	42
3.6	Flash card: fishing	43
3.7	Playing card map: front and back	44
3.8	Playing card: 6 of Wilderness	45
3.9	A straight hand from the *Delta Primer* deck	46
5.1	Catalogue of climatic devices, Jade Eco Park, 2012–2016, Taichung, Taiwan	67
5.2	Cooling climatic devices, Jade Eco Park, 2012–2016, Taichung, Taiwan	68
5.3	Detail of the plan with effect of climatic devices, Jade Eco Park, 2012–2016, Taichung, Taiwan	69
5.4	Functioning stratus cloud device on site, Jade Eco Park, 2012–2016, Taichung, Taiwan	71
5.5	Stratus cloud cooling device on site, Jade Eco Park, 2012–2016, Taichung, Taiwan	71
6.1	Agnes Denes, Isometric Systems in Isotropic Space Map Projections: The Snail, 1979	81
6.2	Dymaxion map of energy slaves, 1945	83
7.1	Sverdlovsk No. 44, the ideal city of the Stalin–early Khrushchev period, 1950–1960s	94
7.2	Sverdlovsk No. 44, the ideal city of the Brezhnev period, 1970–1980s	94
7.3	Sverdlovsk No. 44	95
7.4	Workshop Sverdlovsk No. 44, 2011	95
8.1	Notes on a Notation System, University California, Berkeley, graduate seminar, November–December 1964	106
8.2	Lawrence Halprin, RSVP Cycles, 1969	109
8.3	Portland Open Space Sequence Notes, 1967. Published in Sketchbooks of Lawrence Halprin (1981), p. 61	111
8.4	Portland Open Space Sequence, Tuolumne River, 1962. Published in Sketchbooks of Lawrence Halprin (1981), p. 67	112
8.5	Portland Open Space Sequence, Notated Observation of the High Sierras, 1962. Published in Sketchbooks of Lawrence Halprin (1981), p. 64	113
10.1	Valleys of Lisbon	132
10.2	Evolution of Chelas	134
10.3	Spontaneous *hortas*	135
10.4	*Parque Agricola de Chelas*	136
10.5	Sheds and teepees	137

11.1	Denied access to Paternoster Square, Occupy LSX encamped at the bottom of the steps of St Paul's cathedral in 2011	146
11.2	Paternoster Square, during the Occupy protests of 2011, was fenced off from undesirable protests, people and activities	148
11.3–6	Mierle Laderman Ukeles, *Manifesto for Maintenance Art, 1969 'Proposal for an exhibition: "Care"'*	158
11.7	Co-authored maps, such as Incomplete Cartographies, can entwine contrasting and hidden narratives to establish future design proposals	159
11.8	Research map for *Walking the Elephant* (Dawes and Wall, 2013)	160
11.9	Odd Lots, an exhibition of Gordon Matta-Clark's *Fake Estates*	161
12.1	Bangkok Transit Map with Chao Phraya River	166
12.2	Bang Bua Thong District *khlong* (canal) and community with repurposed canal-side pathway connection	166
12.3	Informal settlement communities dot the landscape in juxtaposition to new luxury residential towers	168
12.4	A CODI-funded *Baan Mankong* project rises from the ground	169
12.5	Community architects, students and residents collaborate over future development plans and strategies	171
12.6	A maturing CODI community exudes ownership, pride and resilience against a challenging marsh landscape	173

CONTRIBUTORS

Ed Wall is the Academic Leader Landscape at the University of Greenwich, London, Visiting Professor at Politecnico di Milano (DiAP) and City of Vienna Visiting Professor 2017 (SKuOR) for urban culture, public space and the future – urban equity and the global agenda. Ed's research focuses on the design and theory of landscapes, public spaces and cities. He is the founding editor of the design research journal *Testing-Ground* (2015). In 2007 Ed established *Project Studio*. Award-winning projects have been published and exhibited widely, including at the Architecture Foundation, Royal Academy, Biennale of Landscape Urbanism, London Festival of Architecture and the Van Alen Institute.

Tim Waterman is senior lecturer and landscape architecture theory coordinator at the University of Greenwich, and a tutor at the Bartlett School of Architecture, UCL. He writes for a wide range of professional and academic publications on the subjects of power, democracy, taste, foodways, and everyday life.

Camillo Boano, is Professor of Urban Design and Critical Theory at the Bartlett Development Planning Unit, University College London (UCL), where he co-directs the MSc in Building and Urban Design in Development. He is also co-director of the UCL Urban Laboratory. Camillo has worked and researched in South America, the Middle East, Eastern Europe and South-East Asia. His research is centred on the encounters between critical theory, urban and architectural design processes and informal and contested urbanisms. He is author, with William Hunter and Caroline Newton of *Contested Urbanism in Dharavi: Writings and Projects for the Resilient City* (2013) and *The Ethics of a Potential Urbanism: Critical Encounters Between Giorgio Agamben and Architecture* (2017).

Paul Cureton is a senior lecturer in Architecture and Unit Leader for Future Cities at the University of Hertfordshire, Member of the Herts UAV Group, Senior Research Fellow in Landscape & Infrastructure at Birmingham City University.

He holds a PhD in Landscape Architecture Representation from the Manchester School of Architecture. Primary research interests include digital and analogue interfaces GIS, UAVs, mapping, modelling and digital fabrication. Recent publications include the monograph *Strategies for Landscape Representation: Digital and Analogue Techniques* (Routledge 2016) and the co-authored governmental working paper *A Visual History of the Future* (Foresight, BIS 2014).

Jill Desimini is a landscape architect and Assistant Professor of Landscape Architecture at the Harvard University Graduate School of Design. Her research focuses on cartographic techniques and design strategies for urban abandonment. She has published articles in numerous journals including *Landscape Journal*, *JoLA* and the *Journal of Urban History*. She is co-author of the book *Cartographic Grounds: Projecting the Landscape Imaginary*.

Ross Exo Adams is Assistant Professor of Architecture and Urban Theory at the College of Design, Iowa State University. His research looks at the historical and political intersection between circulation and urbanization and he has published widely on relations between architectural practice and geography, political and legal theory, political ecology and philosophy. He has taught at the Bartlett School of Architecture at University College London, the Architectural Association, the Berlage Institute in Rotterdam, and the University of Brighton. He is author of *Circulation and Urbanization*, forthcoming from SAGE in the Society & Space series edited by Stuart Elden (2017).

Maria Shéhérazade Giudici is the founder of the publishing and educational platform Black Square. She is the coordinator of the history and theory course at the School of Architecture of the Royal College of Art, and Unit Master of Diploma 14 at the Architectural Association. Before joining the AA, Maria taught at the Berlage Institute and BIArch Barcelona and worked with offices BAU Bucharest, Donis and Dogma. Maria's ongoing research focuses on the relationship between architecture and the construction of subjectivity; her latest publication is *Rituals and Walls: The Politics of Sacred Space* (2016), co-edited with Pier Vittorio Aureli.

William Hunter is a trained architect, urban designer and educator working at the critical crossroads of community design and city planning. From 2009 to 2013 he was a Teaching Fellow at University College London's Bartlett Development Planning Unit where he led urban design studios and field research consultancies including projects in Mumbai, Bangkok, Rome and Medellin. Prior to this, he worked as a designer for Daniel Libeskind, Norman Foster and Thomas Heatherwick. He recently served as a Senior Researcher and Adjunct Professor at the University of Nevada Las Vegas's Downtown Design Center. Current professional practice work includes assisting communities and municipalities across the American Deep South on mitigating the social, physical and ecological impacts of large-scale infrastructure projects.

Jane Hutton is a landscape architect and Assistant Professor at the University of Waterloo, Department of Architecture. Her research looks at the externalities of material practice in landscape architecture, examining linkages between the sources and sites of common building materials. She is currently completing a manuscript for a book titled *Reciprocal Landscapes: Cases in Material Movements* to be published by Routledge in 2017. Hutton is a founding editor of the journal *Scapegoat: Architecture, Landscape, Political Economy*, and is co-editor of issues: 01 *Service*, 02 *Materialism* and 06 *Mexico D.F./NAFTA*.

Katya Larina is an architect and urban designer who received her Master's degree in Landscape Urbanism from the Architectural Association, London. Katya has co-directed a series of independent workshops. She has also been invited as an expert and guest lecturer at Institute Strelka, Architectural Association and the UCL's Bartlett School of Architecture in London. Katya is co-founder of the research and education project U:Lab.spb, which develops tools that are used in the fields of design and analytics of critical urban environments of Russian cities. U:Lab.spb focuses urban strategies that combine knowledge from the fields of sociology, economics, urban planning and ecology to foster the redevelopment of Russian industrial cities and knowledge centres. U:Lab.spb as urban design consultants has been involved in a series of regional development projects and socio-economic strategies for Russian cities, including some research projects on the 'closed cities' phenomenon.

Peg Rawes is Professor in Architecture and Philosophy and Programme Director of the Masters in Architectural History at the Bartlett School of Architecture, UCL. Recent publications include: *Equal By Design* (co-authored with Beth Lord, in collaboration with Lone Star Productions, 2016); 'Housing Biopolitics and Care' in *Critical and Clinical Cartographies*, ed. A. Radman and Heidi Sohn (2017); 'Humane and Inhumane Ratios' in The Architecture Lobby's *Aysmmetric Labors* (2016); *Poetic Biopolitics: Practices of Relation in Architecture and the Arts* (co-ed., 2016); *Relational Architectural Ecologies* (ed., 2013).

Douglas Spencer is the author of *The Architecture of Neoliberalism* (Bloomsbury 2016). He teaches and writes on critical theories of architecture, landscape and urbanism at the Architectural Association and at the University of Westminster, London. A regular contributor to *Radical Philosophy*, he has also written chapters for collections such as *Architecture Against the Post-Political* (ed. Elie Haddad and Nadir Lahiji, Routledge 2014) and *This Thing Called Theory* (ed. Teresa Stoppani, Giorgio Ponzo and George Themistokleous, Routledge 2016), and published numerous essays in journals such *The Journal of Architecture*, *AD*, *AA Files*, *New Geographies*, *Volume* and *Praznine*.

Amy Strecker is Assistant Professor and ERC postdoctoral researcher at the Faculty of Archaeology, Leiden University. Amy obtained her PhD in international law from the European University Institute, Florence, in 2012. Her PhD, which was funded by the Irish Research Council, analysed the protection of landscape as expressed in cultural heritage law, environmental law and human rights. Her current research within the ERC project Nexus1492 focuses on the role of law in confronting the colonial past in the Caribbean, specifically in relation to land rights, cultural heritage and restitution. Amy has been actively involved with the European Networks for the Implementation of the ELC since 2007 and she is currently preparing her monograph to be titled *Landscape Protection in International Law*.

Jane Wolff is an Associate Professor at the Daniels Faculty of Architecture, Landscape, and Design at the University of Toronto. The recipient of two Fulbright scholarships and of research grants and fellowships from the Landscape Architecture Foundation, the Exploratorium, the Graham Foundation and the Harvard Graduate School of Design, Ms Wolff holds a Bachelor's degree in visual and environmental studies from Harvard and Radcliffe Colleges and a Master's degree in landscape architecture from the Harvard Graduate School of Design. In 2006, she was Beatrix Farrand Distinguished Visiting Professor at the University of California, Berkeley.

ACKNOWLEDGEMENTS

Landscape and Agency is the result of many conversations. Our collaboration with Murray Fraser and Douglas Spencer, which began in the Bree Louise pub on London's Euston Street and resulted in the 'Landscape and Critical Agency' symposium at University College London in 2012, is of particular note. We would like to sincerely thank all those who were involved in the symposium, especially the generosity of the speakers, Jill Desimini, Matthew Gandy, Jon Goodbun, Jonathan Hill, Jane Hutton, Douglas Spencer, Lisa Tilder, Jane Wolff and Daniel Zarza.

We are immensely grateful to all the authors who have written chapters for the book. The breadth of discourses contributed from Ross Exo Adams, Jane Hutton, Jane Wolff, Amy Strecker, Maria Giudici, Peg Rawes, Katya Larina, Paul Cureton, Jill Desimini, Camillo Boano, William Hunter and Douglas Spencer have been both daunting and a delight to edit. The process of creating this book has been more than just collecting chapters; it has been a rich and stimulating ongoing conversation. We would especially like to thank Murray Fraser for writing the foreword and Don Mitchell for his stimulating afterword, and of course for their voices in the conversation.

In the five years since the symposium many of our friends and colleagues have inspired and critically read our ideas around landscape and agency. The review of our writing by Christoph Lueder, Lara Rettondini, Oscar Brito and Allan Atlee has been an important part. David Grahame Shane at Columbia University and Antonella Contin at Politecnico di Milano have also offered insightful critiques of our ideas and writing in the course of this book. Suzi Hall and Fran Tonkiss at the Cities Programme at the London School of Economics have greatly informed Ed's understanding of landscape within the context of urban redevelopment. Tim has benefitted from rich conversations with David Haney at the University of Kent and Allen S. Weiss at New York University, while Ruth Catlow and Marc Garrett at Furtherfield have helped relate Tim's discussions about landscape to other large socio-political realms, including the virtual. The thinking that has occurred in the 'Reading the Commons' project there has influenced and broadened our thinking. Included in those discussions have been Anne Bottomley, Joss Hands, Alastair McCapra, Nathan Moore, Christian Nold, Penny Travlou and Ele Carpenter.

As we have progressed these discourses towards this publication we are grateful to Louise Baird-Smith, who during her time at Routledge initially embraced our proposal and more recently, Sade Lee, Grace Harrison and Aoife McGrath whose editorial support has been invaluable.

The initial symposium was only possible due to the generosity of our academic institutions, at University of Westminster, the Architecture Association, Kingston University, Writtle School of Design and the Bartlett School of Architecture, UCL. The Landscape Institute provided sponsorship which, importantly, released us from financial burdens and the need for other fundraising efforts to allow us to focus on working and thinking. We would like also to thank our current academic home, the University of Greenwich Department of Architecture and Landscape, and especially Neil Spiller and Nic Clear for their support.

Finally, we would like to thank our families, especially Kristin and Jason for their immeasurable love and patience as we have taken over many of our evenings talking about the pressing issues of landscape.

FOREWORD

Murray Fraser

This book on *Landscape and Agency* represents the coming together of discussions with friends and colleagues from across a range of universities in London – most notably with Douglas Spencer, Ed Wall and Tim Waterman – to explore critical practices and theories around landscape, particularly in its vital and complex relationships with architecture. More specifically, some five years ago now, after rejoining the Bartlett School of Architecture at University College London (UCL), I was looking to run a challenging event to inject a bit of energy. From this arose the 'Landscape and Critical Agency' symposium, held at UCL in February 2012. The theme was chosen because I realized that the Bartlett, despite its reputation as one of the best architecture schools in the world, did not deal with landscape. The only exception was my colleague Jonathan Hill, a contributor to the original symposium, and whose interest in eighteenth-century Picturesque theory is a key part of his historical and theoretical writings. But his remains a lonely Bartlett voice on the subject of landscape.

In terms of the intentions of our initial 'Landscape and Critical Agency' symposium, I had been involved, while working at my previous institution, the University of Westminster, in running the 'Critical Architecture' conference in November 2004 along with Jonathan Hill and Jane Rendell from the Bartlett, and Mark Dorrian from Edinburgh University. It was followed in late 2007 by a Routledge book of the same name. The aim of that conference/book was to challenge the attempted de-politicization of architecture by the so-called 'post-critical' theorists in the USA. The latter were arguing the truly ridiculous proposition that architecture could not – indeed should not – have any critical or political stance. The 'Critical Architecture' conference set out to oppose them. As such, it proved influential in changing the tone of the debate, mainly by showing the reality that architecture has, and always will have, a political dimension to its discourses and practices.

Also while teaching at Westminster University, I had helped to facilitate the 'Emerging Landscapes' conference in June 2010, as organized by Davide Deriu and colleagues. That too had proved a very stimulating event, and therefore at the Bartlett in early 2012 my idea was, quite simply, to merge these two predecessors through the symposium on 'Landscape and Critical Agency'. Seeking allies in my cause, I turned to Douglas Spencer, one of my doctoral students at Westminster and a contributor to this current book, as well as Ed Wall and Tim Waterman, now both teaching at the University of Greenwich. I knew that Doug, Ed and Tim had each also been exploring critical perspectives of landscape through the work they had presented at the 'Emerging Landscapes' conference. Since collaborating on the 'Landscape and Critical Agency' symposium at UCL, Ed and Tim have graciously taken on the role of editing this current book. I am delighted to see that it has now been expanded to include both a selection of speakers from the symposium and also other writers that are able to present significant challenges to contemporary landscape theory and practice.

This, then, prompts an obvious question: what did the original 'Landscape and Critical Agency' symposium set out to do? At one level, that is easy enough to explain, as we had set out a clear central theme in our call for contributions:

> What *agency* does landscape possess, as a means of territorial organisation and creative production, to engage critically with the conditions that define the collective aspects of our environment?

Having posed this question, however, we also realized that it merely opened up many other issues and problems. Hence, we added in some further prompts to help potential participants. How might we develop a model of landscape that has agency? What place might be accorded to design, to the material and environmental processes of the landscape itself, and to the various users of territory within this model? Given that landscape design cannot ever constitute an autonomous practice, how and where might its agency operate critically? What arguments, tactics and manoeuvres might it need to develop to do so? And so on.

Asking these kinds of questions helped to open up a new approach within landscape discourse, certainly among British and continental European participants. Here we discovered a remarkable contrast between those countries and the situation in North America. In terms of a critical approach to architecture, there is no longer anything of real substance in the USA or Canada. One can of course think of a handful of exceptions: Teddy Cruz, Michael Sorkin, Peggy Deamer, Keller Easterling, Bill Menking and Danny Abramson. But these now only prove the rule, which is that North American architecture, whether based in academia or practice, is apolitical and apathetic. In notable contrast is the field of landscape architecture in the USA and Canada, which has many critically engaged programmes and research centres such as those located at Harvard, University of Toronto and University of Pennsylvania. Conversely, Britain possesses a strong critical architecture tradition in schools at the likes of the Bartlett, Westminster, Sheffield and Newcastle, but has relatively little in terms of critical landscape architecture, with its courses in the subject being more concerned with training professionals than asking larger social and political questions about what they do. The situation is also somewhat similar in continental Europe, which I have to say came as rather a surprise to us when we read the results of our call for contributions to the 'Landscape and Critical Agency' symposium. In contrast, the writings of Douglas Spencer at the Architectural Association and University of Westminster, and of Ed Wall and Tim Waterman at the University of Greenwich, are now very much seeking to change these prevailing practices in Britain and Europe. Collaborating with them has undoubtedly revealed for me a number of fascinating critical discourses around ideas of landscape, along with innovative practices in landscape design, landscape architecture and landscape urbanism.

Hence this fascinating edited book acts as a medium for transatlantic fusion between the more critical landscape design milieu in North America and those in Britain and continental Europe, which both need a real shake up. Likewise, it deliberately aims to address the more abstract theoretical positions in landscape and also actual case studies on the ground. Authors include highly respected practitioners such as Jane Wolff and Jill Desimini in Canada and the USA respectively, as well as wide-ranging theorists like Peg Rawes and Camillo Boano, both colleagues of mine from UCL in London. The array of chapters in the book is lively and challenging. None try to be definitive, but rather hope to provoke those involved in landscape theory and practice to become more politically explicit about what they do – and asking, above all, what are the forms of agency that landscape research and projects might potentially have.

Bartlett School of Architecture, UCL
October 2016

INTRODUCTION

Critical concerns of landscape

Ed Wall and Tim Waterman

In *Landscape and Agency* we explore the capacity for action and change in the interrelations between people and land. We discuss the potential for transforming landscapes and the transformative potential of landscapes to redefine cities and enrich daily lives. We are focused explicitly on landscapes of practice – lived, acted, engaged landscapes. Chapters critically examine landscapes that are formed through a range of interactions across the design, planning, occupation and use of land. We embrace a range of definitions of landscape, from the narrowly scenographic approaches drawn from the field of painting to geographies describing ecological and economic dynamics at territorial scales. The term 'landscape' has not only various contemporary meanings, but has been differently employed through time, and in each era what it has described about our human relationship with land reflects broad shifts in thinking and modes of acting and being. Its earliest European meanings, elaborated with great historical rigour by Kenneth Olwig in his *Landscape, Nature, and the Body Politic* (2002), comprise an explicit relationship between land use (often in the context of the negotiation of the commons), custom, morals, and law. Rather than the largely scenographic and painterly art it would later become, a 'landscape' was first understood as something produced through social and technical changes to the land, the manipulation of the valley basins, coastal shores, rivers and wetlands to provide sustenance, shelter and defence. The meaning of the suffix *-scape* as describing a system or a set of processes and relationships is here evident, and other authors have resurrected this pre-modern and pre-capitalist sense (Jackson, 1984; Bender, 1993; Corner, 1999; Johnson, 2007) including Denis Cosgrove, who acknowledged the resurgence of this definition in his introduction to the second edition of *Social Formation and Symbolic Landscape* (1998). Of course, landscape also plays an important role in providing both an active and passive symbol and a theatrical backdrop for national identity (Daniels, 1993; Matless, 1998; Olwig, 2002). This sense of a collective work need not be tied to nationalism, however, as the work of landscape precedes the construction of the nation-state. Instead, landscape as a collective work over time is implicit, as exemplified by Henri Lefebvre's resonant words, 'We have learned how to perceive the face of our nation on the earth, in the landscape, slowly shaped by centuries of work, of patient, humble gestures' (2008 [1991], 134).

This collectivity is evident in J.B. Jackson's *Discovering the Vernacular Landscape* (1984, 8), that the term 'landscape' was taken in the tenth century to refer to the 'collective aspects of the environment', and this sense of collectivity and interaction lends itself to being conceived in terms that use ecosystem as metaphor. In the last century, an ecological structure of thought has been argued to allow us to conceive of our relationship with landscape in ways that both reach into the history of humanity and which create new frames for thinking about the present and the future that are embodied, emplaced and practised (Bateson, 1972; Code, 2006; Rawes, 2013). This layering-up of meanings allows for a simultaneity of ideas of landscape: that characteristic that consistently calls for the metaphor of the palimpsest to be pressed into service, to the point of cliché. These layers might now be composed of pre-literate embeddedness, medieval ties of custom and obligation, modern scenographic iconographies and their concomitant identities, and now a contemporary ecology of mind which brings them all together in modes of being, acting and thinking. A human environment characterized by such relational action, which reflects on the past to project into the future, is powerfully emerging as landscape's new epistemology. Rather than emerging solely from landscape discourses a series of 'turns' in critical theory have changed landscape's frame. These include the linguistic turn, the cultural turn and the relational, corporeal, spatial, constructivist and ecological turns. These widely linked shifts in critical theory have prompted an examination of agency in landscape, which, in keeping with the post-disciplinary construction of the spatial turn, combine the sociological, the geographical and the historical in equal measure (Soja, 1996, 2010).

Chapters

In *Landscape and Agency* we advocate for the capacity of landscape, as a complex of powerful social, spatial and ecological relations, to empower change, if not also to embody it. While critically reflecting on contemporary theories and practices we aim to highlight the achievements of architecture, landscape architecture and urban design to employ social and ecological knowledge in the rehabilitation of physical sites (see Chapter 3, 'Agency, advocacy, vocabulary: three landscape projects', by Jane Wolff). The continued influences of Landschaftspark Duisburg Nord, Parc de la Villette and Fresh Kills Park on subsequent landscape projects and discourses have endured alongside critical writings on the power of landscape as an ecological and projective medium that can transform cities (see Cosgrove, 1998 [1984]; Corner, 1999; Waldheim, 2006; Reed and Lister, 2014). The chapters of this book have been brought together, in part, due to the persuasiveness of these discourses. However, we are also conscious that many of these design approaches advance economic agendas that tend to situate themselves in the midst of vast, complex, ineffable market systems while assuming a self-organizational or self-regulating equilibrium. Douglas Spencer's chapter on 'Agency and artifice in the environment of neoliberalism' (Chapter 13) directly describes these concerns, as does his timely *The Architecture of Neoliberalism* (2016). We recognize a silence amongst landscape practices and an impotence of landscape, architecture and landscape architecture theory to propose effective address to other contemporary challenges of our environments and societies. We identify that the repeated flooding of cities and the destructive actions of material displacements during urban developments struggle to be contained. We are concerned with the appropriation of landscape – that landscapes as languages, as technologies, as practices and as representations have been co-opted to advance economic advantage despite or against social and environmental concerns. In this context, landscape architectural practices and aesthetics have been actively employed in advancing gentrification, facilitating the displacement of communities and visually framing (and concealing) urban narratives of predation and dispossession that, while obscured, may still be read in urban landscapes by a practised eye. Under the cloak of the acclaim for the successes of landscape projects which, for example, clean up and remake urban waterfronts to open up public access, these new urban landscapes are embraced by landowners and developers (and accepted by municipal planners) as devices to escalate land values, displace undesirable businesses, upturn land uses and attract more affluent residents. In her critique of the development of the Fresh Kills Park on a vast landfill site on New York's Staten Island, Linda Pollak (2002, 59) describes a 'magic disappearing act' which is performed as a veneer of scenographic landscape is employed to conceal the wasted (albeit treated) land below. Landscape, in these terms, becomes a purifying agent remediating the soil as it can also be found to cleanse the contents and beautify the appearance of cities. Following Pollak, many of the chapters are interested in the agency of working landscapes (see Don Mitchell's afterword, 'Landscape's agency'), such as the vast engineered layers of waste being processed below Fresh Kills Park, and they are critical of the agency which scenographic relations use to obscure the knowledge, skill and labour required to create them.

The chapters question prevailing ideas and approaches to landscape with the aim of revealing alternative cultural and ecological relations. We explore how the development of ecological and relational landscape thought and practice can fundamentally change our engagement with people and environments (see Chapter 6, 'Planetary aesthetics', by Peg Rawes). We discuss the development of contemporary landscapes from their recent history in capitalist modernity as a picturesque visual medium to disciplinary developments from landscape gardening and landscape architecture to landscape urbanism. We are concerned with the many ways in which the relationships between the ideas and practices of landscape and the social and subjective formations and material processes are invested with agency. We critically examine the role of landscape in uneven processes of contemporary urban development, environmental debate and political agendas and we ask how these relationships can be analysed and rethought through the dialogic construction of theory and practice. The chapters are focused on the agency of landscape relations, the influence of our environments and our impact on the land through use, occupation, design and development (see Chapter 9,

'Publicity and propriety: democracy and manners in Britain's public landscape', by Tim Waterman). It is through landscape practices (such as landscape architecture and landscape urbanism) that the agency of landscape can be most vividly read: as Ross Exo Adams writes in Chapter 1, 'Landscapes of post-history', 'If there is agency in landscape practices, it is likely grounded in the ontological status of landscape itself.' How the dynamics of landscapes empower individuals and groups to act is a consistent theme through the chapters. As the potential of landscapes are realized through the means of design, community action or political resistance, the vast scope and often unrealized capacity of landscape actions are revealed (see Chapter 11, 'Post-landscape', by Ed Wall). We focus on the agency of landscape relations rather than the capacity for change of individuals. Communities are implicated within these relations as well as the actions and processes of our environments. In Chapter 4 ('The law is at fault? Landscape rights and "agency" in international law'), Amy Strecker explores laws that relate to 'rights *to* landscape' and attempts to establish the 'rights *of* landscape'. Even in these terms, the rights of individuals can be found to be more effectively argued than the latter protection of physical environments through the law. The anthropomorphic sense of agency as empowerment, however, is not one that we find consistently problematic. It can be an aid to making natural processes and forces tangible and legible (see Chapter 2, 'Reciprocal landscapes: material portraits in New York City and elsewhere', by Jane Hutton) as we locate ourselves within our varying environments. The chapters in this book are situated socially, politically, economically, ecologically and legally in a particular nexus of time. They reflect on contemporary design projects that continue to transform ecological conditions (see Chapter 5, 'How to live in a jungle: the (bio) politics of the park as urban model', by Maria Giudici) as well as reconsidering representational approaches that have made claims to empowerment, both ecological and social (see Chapter 8, 'Rhythm, agency, scoring and the city', by Paul Cureton). As old systems struggle to adjust to more recent political and technological change, landscapes can provide a means to adapt. Equally they can remain a relic of past ideals, perhaps illustrating where the agency of nostalgia, or at least of the defence of the status quo, trumps the agency of adaptation and evolution in places (see Chapter 7, 'The closed landscapes of Sverdlovsk-44 and Krasnoyarsk-26', by Katya Larina). The broad geographic range that the chapters address represents the reciprocity of landscape discourses and the global relations which are inherent in contemporary landscape practices. The spaces of and for communities and the representation of social groups in the reconfiguration of landscapes is a core concern in the book (see Chapter 10, 'The power of the incremental: agronomic investment in Lisbon's Chelas Valley', by Jill Desimini, and Chapter 12, 'Activating equitable landscapes and critical design assemblages in Bangkok', by Camillo Boano and William Hunter), that, as many landscape practices also aim to address, provides clues to how future landscapes might be made and occupied.

Landscape and agency

When we began planning the 'Landscape and Critical Agency' symposium (2012), with Murray Fraser and Douglas Spencer, in the conversations that would become the basis for this book we shared a common concern for the need for criticality in landscape practices, design and theory. While troubled that economic agendas and political upheavals were undermining the capacity of landscapes everywhere to support and provide the ground for positive change, we were optimistic that shifting the frames in which we commonly think of and form landscapes could begin to realize more potential in built and lived landscapes. That ecological practices in landscape have advanced is a considerable achievement for which landscape architects can claim a central role. How writers and practitioners engage with the urgency of current landscape relations, to challenge escalating land values, community displacements, continued privatizations and advancing gentrification is a great and complex challenge. *Landscape and Agency* aims to open up these questions, challenge prevailing practices and encouraging action that will engender more just and sustainable future landscape relations. Thus, as we refer to landscape and agency, we describe relationships that are not one of dominion over land, flora and fauna, but rather that we see the work of landscape architecture and of planned development as one of involvement and collective action – and that the work itself is much larger than merely that accomplished within a single project or profession. *Landscape and Agency* highlights how people engage in working landscapes over time – landscapes that have distinctive pasts and tangible, planned for, hoped for and lived futures.

References

Bateson, Gregory (1972) *Steps to an Ecology of Mind*. Chicago, IL: University of Chicago Press.

Bender, Barbara (ed.) (1993) *Landscape: Politics and Perspectives*. Providence, RI and Oxford: Berg Publishers.

Code, Lorraine (2006) *Ecological Thinking: The Politics of Epistemic Location*. Oxford and New York: Oxford University Press.

Corner, James (ed.) (1999) *Recovering Landscape: Essays in Contemporary Landscape Architecture*. New York: Princeton Architectural Press.

Cosgrove, Denis (1998 [1984]) *Social Formation and Symbolic Landscape*. Madison, WI: The University of Wisconsin Press.

Daniels, Stephen (1993) *Fields of Vision: Landscape Imagery and National Identity in England and the United States*. Cambridge: Polity Press.

Jackson, John Brinckerhoff (1984) *Discovering the Vernacular Landscape*. New Haven, CT and London: Yale University Press.

Johnson, Matthew (2007) *Ideas of Landscape*. Malden, MA and Oxford: Blackwell.

Lefebvre, Henri (2008 [1991]) *Critique of Everyday Life*, Volume 1, trans. John Moore. London: Verso.

Matless, David (1998) *Landscape and Englishness*. London: Reaktion Books.

Olwig, Kenneth (2002) *Landscape, Nature, and the Body Politic*. Madison, WI: University of Wisconsin Press.

Ong, Walter J. (1982) *Orality and Literacy: The Technologizing of the Word*. London and New York: Routledge.

Pollak, Linda (2002) 'Sublime Matters: Fresh Kills', in *Praxis 4: Landscapes*: 58–63.

Rawes, Peg (ed.) (2013) *Relational Architectural Ecologies: Architecture, Nature and Subjectivity*. London and New York: Routledge.

Reed, Chris and Lister, Nina-Marie (eds) (2014) *Projective Ecologies*. New York: Actar Publishers.

Soja, Edward (1996) *Thirdspace: Journeys to Los Angeles and Other Real-and-Imagined Places*. Malden, MA and Oxford: Blackwell.
Soja, Edward (2010) 'Six Discourses on the Postmetropolis', in Gary Bridge and Sophie Watson (eds) *The Blackwell City Reader*, 2nd edn. Chichester and Malden, MA: Wiley-Blackwell, pp. 374–381.
Spencer, Douglas (2016) *The Architecture of Neoliberalism: How Contemporary Architecture Became an Instrument of Control and Compliance*. London and New York: Bloomsbury.
Waldheim, Charles (ed.) (2006) *The Landscape Urbanism Reader*. New York: Princeton Architectural Press.

1

LANDSCAPES OF POST-HISTORY

Ross Exo Adams

Supernatural

There is a certain magic we invest in the term 'landscape' today. Like 'nature', 'democracy' or 'communication', no one doubts landscape. As much as we are destroying landscapes each day (and perhaps precisely *because* of this), it has become a signifier around which undeniable truths orbit closely. The fact that landscape as a discursive category intersects with fields from architecture, urbanism and art to geography, political theory, anthropology and philosophy is surely in part a response to a shifting set of concerns that have taken hold under the shadow of climate change and the multitude of phenomena this has brought into public consciousness. Landscape, in turn, is one of the key sites in which a fledgling collective aesthetic is taking shape. Today, it finds itself as a category central to the ways in which discourses are being reshaped, opening up questions of land, geological strata, ecological processes, economies of extraction and production, social and legal divisions, infrastructural connectivity and processes of urbanization. Landscape today appears at once as a consistent background against which contemporary problems obtain visibility and, increasingly, the object occupying the foreground itself.

Landscape is often associated with 'agency'. If there is agency in landscape practices, it is likely grounded in the ontological status of landscape itself. In discussing landscape urbanism in his seminal essay, 'Terra Fluxus', James Corner expounded this capacity of landscape perhaps most succinctly, ascribing four fundamental themes to the then nascent practice of landscape urbanism: landscape urbanism would be a temporally based process, it would work through a medium of surfaces, as a practice, it would be grounded in realism, and it would aim to construct a collective imaginary.[1] While tied to a specific type of landscape practice (landscape urbanism), we can nonetheless see how these principles begin to open up a more general ideological understanding of landscape practices consistent with much of today's ongoing work. Perhaps most fundamentally, unlike architecture, art, literature, music or any other artistic medium, landscape pre-exists its creative becoming: to create landscape, is always to transform it. Kate Orff reminds us that landscape is "both a *frame* and a *solution*".[2] More than in any other creative practice, the ontological status of landscape lends itself to the ways in which practices of modifying it come to be. In other words, to define what landscape is, is also to define the means by which to transform it in practice. So, if we want to question the agency of landscape, we might first consider assessing how landscape is ontologically constructed.

From Corner's essay, we can begin to see how landscape builds itself around a dual agenda: it is, on the one hand, the site of dynamic, horizontal connectivities – the space of forces, flows and processes. As such, designing landscape is an affirmatively non-object-driven practice, but rather a relational 'staging' of systems. Its status, unlike architecture's, is a catalyst of multiple processes – an instigator for an 'ecology of events' to emerge. And indeed, across the discourses of landscape, terms like 'engagement', 'plurality', 'non-hierarchical', 'indeterminate', 'ephemeral' and so on seem to constitute landscape's inherent properties as much as they also designate the basic outlines for practices of transforming it. On the other hand, the contemporary role of landscape in the city makes it a primary site in which to reimagine the contemporary public realm. Thus, all of its inherently non-hierarchical, relational and dynamic capacities are put to work toward constructing a practice equally invested in landscape as a representational medium – the surface on which an emergent symbolism can take root. The two sides of contemporary landscape reveal it to be at once biological and pedagogical; productive and narrative; functionally indeterminate and culturally over-determined.

Archive

We can imagine that contemporary landscape discourses and practices draw a certain influence from discussions around New Materialism and from renewed interest in material cultures. On the one hand, given the trajectory that landscape has inherited from the likes of Corner et al., landscape is endowed with a kind of ontological predisposition toward agency. This sentiment has benefitted in part thanks to thinkers like Jane Bennett and her political ecology of matter,[3] which suggests an agency that dwells in the more-than-human ecology of actors. Landscape, in this sense, much like Corner's version of it, may be seen as a kind of thickened substrate of 'quasi-agents' – "forces with trajectories, propensities, or tendencies of their own".[4] Not only is landscape the medium of naturally existing forces, flows and processes, but the very matter that constitutes landscape itself – the rocks, the soils, the fossils it produces – all add temporal, ecological and geological dimensionality to its 'vitality' – its non-human agency. On the other hand, this agency, if well documented in its materiality, can play a more interpretive role in constituting a kind of historical narrative of human culture – a means to probe the recent and deep past of the human condition in relation to objects extracted from or placed within the landscape. Taken together, these various discursive tendencies instigate a practice in which the rigorous documentation of materials, plants and their unique ecologies can be curated to reveal a kind of social and cultural agency that passes through the material fragments of landscape, entangling the human and non-human worlds in a complex, more-than-human ecology. Landscape, we could say, appears today as a kind of as-found archive of social and cultural history.

If landscape is an archive, then our interventions into it can become the making visible of the richness of its historical evidence; like a well-cut ice core, it must draw us into the past, narrating the unfolding of the human and non-human relations layered into the ways in which landscape now speaks in the present. And if landscape is to be seen as a spatial and material record of the past, it is inevitable that this record will speak of past errors and inherited social systems, recounting – often indirectly – the exploits of capitalism, modernity, imperialism and episodes of human and ecological violence.

Yet here, a surprising thing happens: as much as such material histories may open up questions of politics, by constituting this politics through the various 'ecologies of matter', it often has a counter-political effect: the complexity, violence and injustice that such material cultures of landscape may illuminate often appear ungraspable in the present, either speaking of histories long since past or inviting us to encounter ongoing atrocities such as climate change as comprehendible only through the sublime awe of total, inevitable catastrophe. Either way, when engaging landscape-as-archive, our perception of it seems trapped in one form of contemplation or another. If agency exists in the way the materiality of landscape reveals these histories to us, it all too often comes at the cost of displacing agency from the political realm, suturing it instead to an exclusively material-cultural entanglement curated in a present which, itself, is drained of the political. This sentiment is captured acutely in the announcement of a recent exhibition at the University of Wisconsin, Milwaukee, *Placing the Golden Spike: Landscapes of the Anthropocene*:

> Through photography, sculpture, video animation, film, performance, and participatory events, the exhibition invites us to contemplate how the manipulation of local ecologies and the global exploitation of natural resources will require new ways of living in the 21st century. The exhibition and accompanying programs challenge visitors to recognize the omnipresence of human impact on contemporary landscapes – suggesting that the closer and more carefully we look, the more places we may find to place a golden spike of the Anthropocene.[5]

Why is this? What is it about landscape as a category that forces what is overtly political in content to become passive in its reception? What effect does this have on framing the ways in which we intervene in landscape?

Divide

According to philosopher and geographer Augustin Berque, the modern understanding of landscape – as a theoretical form of knowledge – appeared in Europe in the fourteenth century when Petrarch ascended Mont Ventoux and was moved to reflect on the beauty its vistas presented to him. This, he says, is the moment when landscape "begins to exist for the Europeans".[6] Landscape, here, denotes a subject of thought for which the object (landscape) must exist as something *representable*. Notwithstanding its tautological definition, Berque's is one that places landscape in the realm of the philosophical and the aesthetic. With this in mind, he marks a distinction between landscape *theory* and landscape *thought*, a divide that emerged with the modern construction of the former to the detriment of the latter's more ancient status. This divide allows a rather moralized symmetry to cut through the entire text: landscape thinking is the more primordial, non-Western, non-urban, 'spontaneous' way in which humans have for millennia taken landscape (again, tautologies aside) as an indispensable part of what it means to dwell in time and space. *Landscape thinking* requires an intimate and immediate sensitivity of land, its authentic processes, natural transformations and the social entanglements it constructs across generations. *Landscape theory*, on the other hand, emerges as a reduction of landscape to 'false' representations of itself, seen from an otherwise 'disinterested' gaze looking from the city outward. It is landscape thinking in reverse, where landscape is constructed through a cold logic and becomes the space onto which projections of class structures appear; it is the formation of a rationality born of the artificial, elitist distance in between subject (the 'leisure class') and object (the landscape). He attributes this degraded view of landscape, and its subsequent invention of landscape as theory, to what he broadly calls the 'CMWP', or the 'Classical Modern Western Paradigm', that formed somewhere in the seventeenth century. The source of our contemporary, 'corrupt' fascination with landscape – coinciding with our incessant destruction of it – is, Berque asserts, rooted in the CMWP.

What is curious about this rather moralized hypothesis is how the concept of landscape, which is decidedly modern in origin,[7] dating from the seventeenth century, accords for Berque to a predominantly pre-modern spatial ontology. While certainly not incorrect, since such an ontology is not replaced outright by the modern one (and that the modern/pre-modern 'divide' is itself problematic), Berque's approach has the effect of portraying landscape as a timeless object that, at one point in history, becomes the hapless victim of the violence that the 'CMWP' imposes. In other words, landscape, for Berque, remains a constant; what changes is the way we humans understand it (either as an authentic way of thinking or as an object of a disinterested, elitist and immoral gaze). Such a reading not only plays to an essentialist depiction of landscape, ignoring the political histories that helped give birth to the concept as we've inherited it, but, more crucially, it overlooks how landscape itself has also come to serve as a fundamental *technology* in the constitution of modern politics (what Berque might call 'CMWP').

Technology

Indeed, what is all too often left out of contemporary landscape discourses is the deeply political history of landscape: the modern history of this category is as much a history of a 'disinterested' aesthetic as it is a history of modern statecraft, and it is not by chance that the emergence of landscape coincides with that of *territory*.[8]

Antoine Picon makes this co-production explicit, locating an origin of the modern concept of landscape in seventeenth-century France as a direct consequence of the technological rationalization of its territory.⁹ Emerging as a concept at the confluence of cartography, geography, politics, economy and gardening, landscape for Picon answered to the mounting demands for creating a space that could be scientifically measurable, economically calculable and controllable as a technology. In the wake of the great epistemological crises that rocked the sixteenth and early seventeenth centuries, shaking apart the remnants of the 'pre-modern' world in Europe, a new, urgent set of ideas occupied thinkers in their desperate attempts to impose order and stability in a world that suddenly seemed devoid of any. Turning away from a theologically ordained world to one organized by geometry and scientific reason, landscape, under the administration of Jean-Baptiste Colbert (minister of finances under Louis XIV), would be conceived as a geometrically disciplined space of controlled *carrefours*, boulevards and *étoiles* of circulation for the wealth and resources that would increasingly constitute the power of the state. Landscape presented itself to modern sovereign power as a space whose natural variation suddenly appeared empty of any inherent significance, and in turn opened itself up as a *tableau* of indifferent differences – a quantitative space available to the rational calculations and organizations of *Raison d'État*.¹⁰

From the treatises of state theory of Giovanni Botero to those of Thomas Hobbes to the policies and programs of Colbert, it is landscape that plays an increasingly central role in the development of the early modern state and its new forms of power. Confirmation of this comes in part in the fact that the experiments carried out by the great French landscape gardener, André Le Nôtre, would singularly shape a century of infrastructural work executed by the *Corps du génie*, and later the *École des ponts et chaussées*, the state engineers who generalized practices of landscape, conceiving it in turn as a technological space measured by triangulation and made systematic through standardized roads, bridges, canals and tunnels. Landscape under this new political rationality would become a space whose composition would be captured in evermore precise cartographic representations and whose expanses would be split by heavily fortified borders constructed by the military engineering of Sébastien Le Prestre de Vauban. In other words, landscape, as a technology of territory, results when the space of the modern state is transformed both physically and epistemologically into a controlled space of circulation, extraction and resource distribution, governed by an increasingly infrastructural and economic form of knowledge. This knowledge by which to (re)produce landscape was not only firmly rooted in the core of the modern state: it was the means by which the state discovered its vast new powers.

Yet landscape is also a contested category. By the latter half of the eighteenth century, with the growing confrontation between the European state and its disenfranchised subjects, landscape again played a central role in the political reforms that rippled through the state, at once strengthening it as a modern paradigm around an evermore economic form of power, while at the same time laying the groundwork for its revolutionary transformation. Physiocracy would play a crucial role in this, advancing some of the first counter-state, macro-economic theories that shared a simple claim that all wealth has its origin in nature – in the landscape. Although the physiocrats' influence in shaping policy was somewhat limited, their work would resonate more profoundly in social and cultural debates and subsequent economic discourses, and in particular with the engineers of the *École des ponts et chaussées*. Indeed, both the physiocrats and the engineers shared landscape as a common discursive object, allowing the adoption of the physiocratic tenets to the engineers' practices of 'improvement': if the origin of wealth was in the landscape, and the source of its growth in its circulation, the physiocratic doctrine gave the clear outline for which the engineers' work could be constituted. It was no longer a matter, as it was for architects, of imitating nature, but rather of mapping it in order to 'perfect' it.[11] The drawings made by the engineers aimed to represent nature in its exactitude and totality. Like Francois Quesnay's *Tableau économique*, these maps would become catalogues of nature itself.[12] 'Reading nature' meant creating a cartographic register of all productive lands, minerals and resources distributed throughout the land. The landscape appeared as an inventory of resources available to the state through its technical arm of engineers. Nature (as resources) and human artifice coincided in this category to form a self-justifying instrument of state-administered production. For wealth to grow, according to Richard Cantillon,[13] it had to be set into motion. Thus, mapping nature could lead to 'perfecting' it through systems of infrastructural communication, which is precisely how the work of the engineers came to correspond so tightly with the thought of the physiocrats.

Here we see how the language surrounding landscape, depicting it as an eternal source of natural truths, would assist in deliberately covering over the deeply political ambitions in the work of the physiocrats and the engineers they influenced. Their intervention, brief as it was, presented *as a necessity* the opening of state trade to a 'world market' based on *laissez faire* policies. The predominant view of nature they adopted saw freedom in economic terms, as the freedom of circulation, which they justified on the grounds of the natural, self-regulating mechanisms inherent to commerce, since commerce itself was nothing if not an expression of nature. Quesnay went furthest in this sense with his development of the '*Droit naturel*', an immutable law which purported that all human actions within the course of physical events regulate themselves according to 'moral law'.[14] Not only did landscape as a rhetorical device help to naturalize absolute political power, but it would provide the ideological cover under which a political reformism could take place, as Foucault has famously shown, in which an emergent diagram of biopower could take refuge,[15] concealed even from its protagonists behind a moralized, enlightened cloak of apolitical critique.[16]

Landscape, as both philosophical-aesthetic object of contemplation and the techno-political medium of territory, is the site of a great experimentation of a new, modern instrumental reason, a politics of prognosis and calculation, the medium of a grand, artificial machine, and the source of its biological reconstitution in the nineteenth century. It is a concept at the core of modern politics and modern power eternally capable of doubling back on itself as a self-evident category of natural, 'apolitical' truth. Like many others born of its time, it is a notion bound far more to conceptions of calculation, control and instrumental reason, than it denotes the philosophical-aesthetic site of wonder that seems to occupy such a privileged position in contemporary architectural, art and urban discourses.

Post-history

It may be the enduring quality of modern landscape to continually double as both a surface onto which we project social and political orders and the source of innumerable truths that, in turn, vindicate them. Much of the language we use today to speak about landscape seems to rehearse a similar motif in which it has been understood for three centuries: both a site of truth and the object of endless struggles over the ability to transform it; a category which opens itself to political instrumentality while constantly retreating to re-present itself through the narrow lens of eternal philosophical contemplation. Yet there is something equally as present in the contemporary moment that both Vilém Flusser and Peter Sloterdijk have referred to as 'post-history'[17] which, today, allows us to speak of landscape as an archive rich in political history, while being completely unable to comprehend its political status in the present. Writing in the early 1980s, Flusser asserts that, following Auschwitz, Western culture has fully realized its complete objectification of human life and has now entered into 'post-history' – a cybernetic world in which programming life becomes the only objective and the annihilation of life, the only outcome. Building on Flusser's ideas, Sloterdijk's more recent interrogation of contemporary capitalism as 'world interior' traces the history of globalization across millennia, whose contemporary outcome in neoliberal capitalism has produced a world space of interiorized comfort and security in which both politics and history have expired and space has been reduced to the fiction of its complete techno-economic scalability.[18] In a crystal palace that appears without an exterior, 'post-history' assembles itself as the creation of witnesses to a history (and thus a politics) that remains out of reach. Captured in popular tropes of 'awareness' or in the diligence of mapping and pedagogy of 'making visible', agency in the post-historical present appears as the transcoding of events, histories and politics into narrated communication – the controlled traffic of information: design that bears a clearly post-historical agency helps to displace politics in its seemingly endless fascination with producing a spatial-material lexicon of good intentions.

If landscape today presents itself as an archive of social and cultural history, its status *as an archive* reveals its post-historical constitution since it presses the politics that its materials 'speak' of immediately into the aseptic frame of contemplative aesthetic consumption. We consume these politics because they are, as such, impossible to engage otherwise. Indeed, such a contemplative understanding of landscape invites us to comprehend our present by innocently cataloging the histories handed down to us. Landscape thus becomes the site on which a perception of a *world bequeathed*, yet never fully belonging to us, forms – a world whose 'objective' innocence ironically helps to deprive it of vibrancy. Through this neo-realist, found-object past, an accidental modernism emerges in the implied, yet indirect, rejection of history as anything other than an archive of the evidence of past wrongs. This quasi-modernism constructs its break with the past not around a particular agenda to propose new and politicized imaginaries of worlds to come, but in trading instead on its utopian innocence through which it discovered itself in the first place, giving way to a practice of radical pragmatism that cleverly avoids the possibility of constructing something we might one day call 'history'. It is here where landscape-as-archive converts itself instantly into landscape-as-program, mobilizing the past it made visible as a post-historical, apolitical system of relational management and the celebration of complex human and non-human entanglements. Yet precisely by avoiding the perils of 'history', such a pragmatism remains unable to comprehend the very problems that its inquiries perpetually unearth: the political ecology of capitalism, the planetary politics of empire, the biopolitics of infrastructure, the violence of urbanization and of course, the long, ongoing history of landscape itself as a political technology. It is perhaps this that motivates architects like Kate Orff to imagine that the politics of global capitalism that produced climate change are somehow best confronted by up-scaling of cybernetic behavioral modification, while never capturing the political implications of such a proposal.[19]

What of landscape's representational capacity? While such landscapes of post-history may not be able to effectively engage the politics they uncover, this does not mean that landscape has become an apolitical category. Far from it. Landscape across discourses has become a prominent medium in which we could say a new monumentality of post-history has begun to take shape: memorials to past atrocities, gardens of hope and peace, monuments of the victims of past (historical) politics, an iconography of previous ecological disasters . . . Contemporary landscape is a strange brew of the pragmatic with the semiotic, the scientific with the allegorical, the infrastructural with the representational. It suggests a practice as much engaged with the biological realism of plants, soils and hydrological infrastructure as it is with constructing a new collective symbolism.

However, while its symbolism may seem to speak to the causes it is called upon to address (war, violence, destruction, etc.), it is perhaps not here where this collective imaginary is constituted. Indeed, the symbolic gestures in many landscape practices appear as markers for its otherwise overtly pragmatic engagement with plants, systems and infrastructures. It is as if the symbolism it often employs to address historical events can only resonate in a post-historical world through the persistence and immediacy of landscape-as-technology: landscape answers to impossibility of history with well-meaning adjectives of systems: *dynamic, relational, productive, flexible, adaptive, open, contextual, inclusive*, etc. At a collective level, landscape congeals in a kind of *aesthetic of the horizontal*, which today serves more as a pedagogical medium that helps bind the status of infrastructure to collective socio-cultural understandings of the world. It is the milieu that, in its intimate connection to a botanical nature, trades on its ability to sustain life as a purely biological category – a life emptied of its historical and political consistency and thus also its subjective agency. Landscape, as such, risks becoming a medium on which an accidental iconography of the biopolitical-cybernetic present will take root.

Command

If the practice of landscape today is trapped in the perpetual curation of its archival present, how can we reconcile its status as, at once, bio- and geopolitically entangled while also drained of its ability to promote a meaningful political agency? A clue to this may come from the notion of *archive* itself, a term whose root – *arkhein* – means

Landscapes of post-history **15**

both to begin and to rule; the archive as *source and command*. Practices that employ a deliberately political use of an archive to understand landscape reveal a far different way in which landscape, instead of foreclosing the political to a bygone era, relentlessly reconstitute it in the immediate present. The artist/activist collective *World of Matter*, for example, see landscape as just such an archive whose contents never stand as a source of contemplation, but are mobilized to force open a host of political debates (Figure 1.1). By understanding the landscapes of materials, resources, infrastructures and lives they examine as eminently political ecologies, the archives they constitute are not only a means to frame a landscape politically, but, in their very presentation, they deliberately outline an activism by which to achieve a certain outcome. Similarly, the Forensic Architecture project led by Eyal Weizman[20] situates landscape as an overtly political category, thus denaturalizing

FIGURE 1.1 Stills from *Black Sea Files* by Ursula Biemann (2005). This project follows the construction of the BTC Oil Pipeline connecting the vast reserves of oil in the Caspian Basin to the global, sea-based network of its circulation. Biemann situates crude oil as an abstract resource in the larger political and social ecologies constituted by the material construction of this massive corridor, marking relations between the subjects of political struggles it produces (oil workers, sex workers, farmers) with the spaces and processes of its construction. Together, they depict the entanglements from which new forms of activism have emerged. Biemann is a core member of *World of Matter*. Image courtesy of the artist.

FIGURE 1.2 *Now*, video installation by Chantal Akerman from a show of the same title of her work in Ambika P3, London (2015)

it, allowing it in turn to enter into a public, political, legal forum to ground the specificities of a given event and marshal them toward a specific outcome. Lastly, the late artist Chantal Akerman's recent installation *Now* (2015) takes landscape as its primary object, capturing it in five, multi-channel, symmetrically arrayed video projections of unnamed, yet contested landscapes in the Middle East, shot seemingly from quickly moving vehicles (Figure 1.2). A rumbling soundtrack of movement, gunshots, explosions and distressed voices accompanies the installation, at the centre of which a floor projection depicts an image vaguely reminiscent of a bed cover, juxtaposing the withdrawal afforded by domesticity with the violence of human displacement and conflict. Landscape here becomes a device that speaks not of itself, but of an immanent and ominous violence, drawing the viewer into the immediacy of ongoing warfare. It becomes a signature not of the truths we associate with a moralized 'nature' and its apparent loss, but of desperation and urgency of conflict in the immediate present. All of which focus on landscape only to destabilize it from its philosophico-aesthetic status of distanced contemplation.

In all of these cases, landscape is never the neutral bearer of a detached history, and thus always carries the inscription of a discursive and political position. Agency appears by allowing landscape to remain a tool (an archive) of power and a stage of human conflict. Its objects, materials and relations are then able to reveal a field of political positions. These practices offer one kind of possibility for landscape to draw from its inherently political consistency in the formation of practices of vibrant activism. Yet surely another kind of agency must engage landscape as the source of a new political imaginary – one not only capable of resisting dominant political forces, but of proposing radically new realities. Why remain so timidly fixated on preserving the conditions of the present (sustainability, resilience, etc.),[21] when it is in our very capacity to imagine the many worlds that exist within and beyond the anti-political landscapes of post-history?

Acknowledgement

Special thanks to Ursula Biemann for her thoughtful comments and discussions and for supplying the original stills used here.

Notes

1 Corner, James (2006) 'Terra Fluxus', in Charles Waldheim (ed.) *The Landscape Urbanism Reader*. Princeton, NJ: Princeton Architectural Press, pp. 20–33.
2 Orff, Kate (2016) *Toward an Ecological Urbanism*. New York: Monacelli Press, p. 7.
3 Bennet, Jane (2010) *Vibrant Matter: A Political Ecology of Things*. Durham, NC: Duke University Press.
4 Ibid., p. viii.
5 An exhibition held at the University of Wisconsin, Milwaukee from 26 March to 13 June 2015. See https://uwm.edu/inova/placing-the-golden-spike-landscapes-of-the-anthropocene/ (accessed 20 February 2016).
6 Berque, Augustin (2013) *Thinking through Landscape*. London: Routledge, p. 2.
7 This point is made clear in Olwig, Kenneth R. (2002) *Landscape, Nature, and the Body Politic: From Britain's Renaissance to America's New World*. Madison, WI: University of Wisconsin Press.
8 For more on the notion of territory, see Elden, Stuart (2013) *The Birth of Territory*. Chicago, IL: University of Chicago Press.
9 Picon, Antoine (2009) *French Architects and Engineers in the Age of Enlightenment*. Cambridge: Cambridge University Press, pp. 100–101.
10 On the notion of *Raison d'État*, see Foucault, Michel (2009) *Security, Territory, Population: Lectures at the Collège de France, 1977–1978*, ed. Michel Senellart, François Ewald, Alessandro Fontana and Arnold I. Davidson, trans. Graham Burchell. London: Palgrave Macmillan.
11 Picon, *French Architects and Engineers in the Age of Enlightenment*, p. 109.
12 Ibid., p. 217.
13 See Cantillon, Richard (c. 1730) *Essai sur la nature du commerce en général*. Paris.
14 Schumpeter, Joseph (1954) *History of Economic Analysis*. New York: Oxford University Press, p. 229.
15 See lecture 2 (18 January 1978) of Foucault, *Security, Territory, Population: Lectures at the Collège de France*.
16 See Koselleck, Reinhart (1988) *Critique and Crisis: Enlightenment and the Pathogenesis of Modern Society*. Cambridge, MA: The MIT Press.
17 Sloterdijk likely builds this concept from the work of the late Vilém Flusser. See Flusser, Vilém (2013) *Post-History*, trans. Rodrigo Maltez Novaes. Minneapolis, MN: Univocal.
18 See Löwenhaupt Tsing, Anna (2012) 'On Nonscalability: The Living World Is Not Amenable to Precision-Nested Scales', *Common Knowledge* 18(3), pp. 505–524.
19 See Introduction in Orff, *Toward an Ecological Urbanism*.
20 Forensic Architecture is a research agency led by Eyal Weizman based out of Goldsmiths College, University of London (www.forensic-architecture.org).
21 Adams, Ross E. (2010) 'Longing for a Greener Present', *Radical Philosophy* 163(September/October), pp. 2–7.

2
RECIPROCAL LANDSCAPES

Material portraits in New York City and elsewhere

Jane Hutton

Originally published in *Journal of Landscape Architecture* 8(1), pp. 40–47.

In the twenty-first century, the rate of material movement from construction and agricultural activity has surpassed, by an order of ten, that of geological processes (Wilkinson, 2005, 161). Designers participate in this monumental shifting, reorganization and recycling of materials around the globe, the great majority of which is bound as urban parks, buildings and highways.[1] These accumulated urban stocks produce at once ecological (the material exchanges produced through construction), economic (the trade made possible through infrastructural networks) and social (the discourse enabled through the public commons) conditions in situ. At the same time, the construction materials that designers specify are implicated in the ecological, economic and social relations of their own extraction, production and re-use. Materials in landscape architecture are physical fragments of remote quarries, factories and forests and their production is responsible for landscape transformation elsewhere. This reality is abstracted and concealed through the commodity form and is an overlooked, yet critical, consideration for the discipline of landscape architecture today.[2] This chapter argues that the study of paired landscapes of production and consumption generates a spatial framework for examining the social and ecological relations of their material exchange.

While much of the focus on materials in landscape architecture has been dedicated to their role in construction assemblies, a range of recent texts have explored the dynamics of materials beyond a single state. These writings address change due to temporal processes such as weathering,[3] dynamic systems involving planting and innovative materials[4] and include references for evaluating the environmental impacts of site materials including the Sustainable Sites Initiative.[5] Evaluation criteria for such systems foreground the cyclical nature of materials used in design, examining multiple lifecycle states and considering issues related to the inputs and outputs of production and disposal. At the same time, they are necessarily reductive in order to produce generalizable principles, further reifying materials into products.

Four precedents for describing discrete relationships between remote landscapes, people and things, are cited here as methodological prompts. In a series of Non-Sites, Robert Smithson assembled quarry materials next to representations of the sites from which they were gathered in the same gallery space. Through the displacement of rocks and sand between quarry and gallery, this work challenged viewers to consider the 'double reflective void occurring between two places and two ways of looking' (Hobbs, 1981, 111). In a 1990 text, David Harvey challenged geography students to trace their last meal through the places, objects and hands that produced them (Harvey, 1990, 22). These analyses of everyday landscapes and objects provide insight into how, as Harvey writes, 'the rules of capital circulation and accumulation . . . get tangibly expressed and actively re-shaped through socio-ecological processes' (Harvey, 2006, 78). Elaine Hartwick's geo-materialist approach to analysing commodity chains links stories of producers and consumers as a crucial political project for radical geography (Hartwick, 1998, 2000). Finally, Latour's Actor-Network-Theory (2005) and William Cronon's tome on the development of Chicago and the Great West (1991) foreground the myriad human and non-human agents (including abiotic materials and sites) involved in the formulation of landscapes. This chapter draws from Smithson's practice of juxtaposing the representations of production sites and designed sites, Harvey's challenge to trace an everyday object in order to de-objectify it, Hartwick's narration of linkages between production and consumption, and Latour and Cronon's shared insistence on the relationship between multivariate actors.

Three cases are examined, each consisting of two sites tethered by the movement of a particular building material from one to the other. The cases represent significant instances of landscape construction in New York City over 150 years in relation to distinct economic periods. The first looks at granite from coastal Maine used in the southern reservoir house in Central Park during the prime of industrial capitalism of the late nineteenth century. The second case links Ambridge, Pennsylvania – the company town of the American Bridge Co. – with structural steel used in Riverside Park during the Keynesian New Deal programs of the 1930s. The third case looks at *Tabebuia sp.* (ipê) lumber from northern Brazil used in the first construction phase of the High Line, reflecting relations of production during the current neoliberal period. Far from a comprehensive history of the material practices of the past century and a half, these three cases instead present portraits. These portraits examine a physical link between two places: an emblematic landscape architecture project and an associated quarry, factory or forest.

Fox Islands, Maine | Central Park, New York City, 1863

The archipelago of the Fox Islands in Penobscot Bay, Maine is underlain with a massive body of pink granite. In an 1880 report, geologist Nathanial Shaler remarked that the stone of Penobscot Bay 'opens easily, having the peculiar inchoate joints that are such striking features in the syenite or granite of New England . . . the lines of

FIGURE 2.1 Sands Quarry in Vinalhaven, Maine, owned by the Bodwell Granite Co. US *Geological Survey Bulletin* 313, p. 1907. US Geological Survey, Department of the Interior/USGS. Photo by T.N. Dale.

weakness in the rocks are not so numerous as to make the quarried masses too small for use' (Rich, 1892, 748). The granite's workability, smooth grey coloration and desirable polish, as well as the promise of exceptionally large blocks, made Penobscot Bay granite desirable for a range of building and paving applications (Figure 2.1).

The string of small islands provided an interminable shoreline with quarries so close to ports that cut rock could be loaded directly onto sloops headed for Boston, New York, Philadelphia and Washington, DC (Merrill, 1891, 184). By the turn of the nineteenth century, the state of Maine produced more granite than any other; profitable government contracts for federal buildings, post offices, bridges and new paved boulevards through the urbanizing capital cities of the states had established a fluctuating workforce (Grindle, 1976, xi). The largest of the Fox Islands, Vinalhaven, nearly doubled in population from 1850 to 1880, the period that saw massive granite blocks sailing south to be stacked at the East River Bridge, DC's State Department Building and erected as columns at St John the Divine in New York (Grindle, 1976, 51).

In 1862, the Board of Commissioners of the Central Park (BCCP) reported the completion of nearly all ornamental bridges, the transverse roads at 86th and 97th Streets and final grading in the reservoir. Newly contoured land, assembled bridges and planted woods were achieved through a massive reorganization of existing topsoil, earth, rock and vegetation from the inhabited land that was cleared to make way for the park (BCCP, 1861, 9). Of those materials brought in from off-site, the farthest flung elements were small or singular, such as the encaustic tiles in the Bethesda Terrace roof, made by the Minton Company from Stoke-on-Trent,

England, and bronze sculptures cast in the foundries of Paris, Munich and Barcelona. The majority of materials, however, came from New York or neighbouring states. Various masonry block types were transported by boat and train from upstate, including variegated gneiss from Westchester County and greywacke from Hudson Valley, and from out of state, including yellow and white brick from Milwaukee, sandstone from New Brunswick and granite from coastal Maine (BCCP, 1861, 86).

As the Central Park reservoir was first filled with water, 700 km to the northeast '60 to 120 men and ten pair of oxen' were cutting stone in Spruce Head Quarry of the Fox Islands for the reservoir's gatehouses (Grindle, 1977, 7) (see Figure 2.2). Successes with other large New York contracts led to the purchase of Spruce Head Quarry by Bodwell Granite Company, ringleader of the Granite Ring. So-called 'Fifteen Per Cent Contracts' on government work guaranteed a hefty profit for quarry owners. In the face of Bodwell's success, and the unstable nature of unregulated work contracts, hazardous work environments and dust inhalation due to new machinery, workers began organizing to form a union, meeting first on Clark Island in 1877. Due to fluctuations within the larger economy and the industry itself, their efforts to organize were not easy. By 1878, Bodwell ordered the dismissal of thirty major union members on its payroll, provoking a walkout and eventual strike (Grindle, 1976, 58) (see Figure 2.3).

FIGURE 2.2 South Gate House, Central Park, 1891 [neg #62683]. Collection of the New York Historical Society.

FIGURE 2.3 Granite cutters from the Bodwell Granite Co. marking the government contract (1872–1888) for the State, War, and Navy Building in Washington, DC. J.P. Armbrust Collection, courtesy of the Vinalhaven Historical Society.

To protect New York cutters from competition with 'cheap labour' in Maine, the 1894 Tobin Law stated that all stone used for New York municipal or state projects would be processed on-site. When the New York City Board of Public Works ruled that this law included paving blocks, Bodwell fired all seventy-five of its Vinalhaven cutters (Grindle, 1976, 88). Later the same year, when the law was amended, Vinalhaven's industry entered a new boom time, sending thousands of tons of blocks to Portland, Boston and New York (Grindle, 1976, 89). Fluctuating demand for masonry construction in urbanizing centres alongside the advent of concrete and steel construction would continue over the next twenty years, producing boom–bust conditions and precarious employment for the granite cutters of the Fox Islands. The Bodwell Company sold out in 1919, and although quarrying continued on the islands sporadically until the end of the Second World War, this closure was symbolic of the end of peak granite extraction in Maine. The eventual demise of the granite industry in the Fox Islands, driven by competition from cheaper granite in international markets, left behind an archipelago of abandoned quarries. Today these quarries have been reclaimed as swimming holes as part of the local landscape of tourism.

Ambridge, Pennsylvania | Riverside Park, New York City, 1937

The convergence of the Allegheny and Monongahela Rivers carve a 40-degree angled landmass – Pittsburgh's Point – and spawn the Ohio River. Pittsburgh's early foundries and mills aggregated around the Point, but development spread rapidly along the three rivers' shores at the turn of the twentieth century. Large, sloped, meandering river sites were optimal for new open-hearth plants that required more riverfront land. Down

FIGURE 2.4 Houses and steel mills in Ambridge, Pennsylvania, 1938. Photographer: Arthur Rothstein (1915–1985). Library of Congress, Prints & Photographs Division, FSA/OWI Collection [LC–USF34–026531-D]

the Ohio River, the American Bridge Company plant was just one of some 140 riverfront iron and steel works within a 30-km radius of the Point in 1906 (Warren, 1973, 134). The company purchased a town formerly known as 'Economy', which had been settled in 1824 by the Harmony Society – a German-Christian pietist community – to establish Ambridge, its own namesake (Slater, 2008, 7). Ambridge rapidly developed Economy's farmland with a grid of streets, workers' housing and the world's largest structural steel fabrication plant at the time – 57 ha – along the riverfront (Figure 2.4).

The American Bridge Company, formed through J.P. Morgan's consolidation of twenty-eight steel manufacturers in 1900, was soon after brought into the United States Steel Corporation – the world's first $1 billion company. The population of Ambridge quadrupled between 1910 and 1929, as the company expanded, other industrial giants arrived and steel workers emigrated from Europe (Slater, 2008, 41). Geographic proximity to large urban centres, connectivity by rail, and changing sources for coke and iron, underwrote the successes of the steel industry. In 1928, for example, some 517,000 metric tons, or nearly a third of the nation's total steel orders, landed in New York City (Warren, 1973, 180).

The Stock Market Crash of 1929 halted operations in Ambridge, with sporadic one- and two-day workweeks supported by few contracts (Slater, 2008, 61). In 1933, the *Chicago Daily Tribune* reported a partial contract for the San Francisco

Bay Bridge and four 150-m steel towers reignited operations at the Ambridge and Gary, Indiana American Bridge Company plants. These and other public works contracts supported through New Deal legislation, and later contracts related to the Second World War, were to become the company's mainstays.

The National Industrial Recovery Act of 1933, signed as part of Roosevelt's New Deal legislation authorized governmental regulation of industry, and protected collective bargaining rights for unions. This stimulated steelworker efforts to unionize and in the same year labour protests in Ambridge resulted in violent clashes at the Spang, Chalfant tubing plant, and the closure of four other plants (*Wall Street Journal*, 1933, 9) (see Figure 2.5). Widespread organizing led to the formation of the Steel Workers Organizing Committee and its 1937 collective bargaining agreement with the United States Steel Corporation, American Bridge's parent company.

In New York City, the Public Works Administration, also part of the New Deal, was responsible for the construction of major infrastructure projects that consumed massive quantities of steel. Many of these projects were focused on routes out of the city or 'new avenues of escape from the island of steel and stone' (Vogel, 1932, A24). The West Side Improvement, finally implemented in 1937 under Robert Moses's lead, introduced the elevated rail High Line in Lower West Manhattan and linked the West Side Elevated Highway with the Henry Hudson Parkway running from 72nd

FIGURE 2.5 Armed deputy sheriffs confronting picketers at the Spang, Chalfant Seamless Tube Company, Ambridge, Pennsylvania. Photograph published by World Wide Photos, 1933. Library of Congress, Prints & Photographs Division [LC-USZ62-26197]

Street through Riverside Park up to Van Cortland Park. Moses's skilful managerial and spatial entanglement of highway, rail and park budgets and agencies mobilized construction funding and emergency relief labour (Caro, 1975, 535). A bill signed by Senator John Buckley authorized funding for relief workers to begin roofing the New York Central Tracks in Riverside Park. 'West Side Project Ready to Hire 4,000: Needy to Get the Jobs', a *New York Times* headline proclaimed (1934, 1).

Olmsted and Vaux's plan for Riverside Park, realized in 1900, negotiated the steep slopes between Riverside Drive and the New York Central rail tracks and the Hudson River to the west. In the years that followed, the expanding rail infrastructure and associated structures brought 'smoking locomotives and sometimes odorous live freight' adjacent to what had become an upper-class neighbourhood (Sweeny, 1937, 14). The land between the rail tracks and the Hudson River was inaccessible and increasingly filled with railroad structures and used as a dump. While countless proposals to reduce the railroad's impacts were made, the covering of the rail tracks with a seamless landscaped surface and the extension of the shoreline with fill became the keystone of Moses's West Side Improvement. The plan promised a landscape 'ungashed by railway cuts and no longer disgracefully fringed with railroad freight yards and unsightly dumps' (*New York Times*, 1936a, E10).

FIGURE 2.6 Riverside Park at 82nd Street under construction, showing steel frames installed over the New York Central Railroad tracks, 1936. Photographer: Samuel H. Gottscho (1875–1971). Milstein Division of United States History, Local History & Genealogy, The New York Public Library, Astor, Lenox and Tilden Foundations.

The roofing of the rail tracks, smoothing of the grade between Riverside Drive and the Hudson River, and integration of parkway and promenade, produced 53 new hectares of parkland valued at $23,760,000. In the promotional document published at the park's opening in 1937, a photograph of the rail tunnel with thin beams of light streaming in is labelled 'Above these covered tracks are grass, trees, and sunlight' (Sweeny, 1937, 49). While appearing as a continuous landmass planted with trees and shrubs, this new sinuous landscape required 45,000 metric tons of structural steel. Rail tracks were covered with frames of 23 m-long riveted steel girders spanning two piers (*Historic American Engineering Record*, 2006, 65). By January 1936, 'literally acres of steel and concrete' had been installed (*New York Times*, 1936b, xx6) and an additional 5,400 metric tons of structural steel had been ordered from the American Bridge Company to cover New York Central's tracks (*Wall Street Journal*, 1936, 3) (see Figure 2.6).

The renovations to Riverside Park wielded New Deal funding and labour, the sectional arts of landscape architecture and the structural capacity of steel to choreograph rail, pedestrian and vehicle movement into a highly constructed, connective landscape. Rail freight from inland manufacturing regions entered the city under the turfed surface of Riverside Park and along the West Side. Above grade, the Henry Hudson Parkway was a new connective typology – a privileged means for escaping the difficulties of the city and consuming the exurban as a landscape of leisure (Gandy, 2003, 125). Even at a small scale, however, West Side amenities were unevenly connected. Between 125th and 155th Streets adjacent to Harlem, the tracks remained uncovered and no investment in parkland was made. African American populations continued to be subjected to the smells and 'never-ending clanking' of the railroad, still open to the air (Caro, 1975, 557).

American Bridge's Ambridge operations were shut down in 1984 as competition from offshore steel production mounted. In 1988, American Bridge was sold to a Taiwanese company, who still maintains the company with a Coraopolis, PA headquarters (Gaynor, 2000). Ambridge's population today is close to that of 1910. Its industrial heritage is most prominently celebrated in the form of Old Economy Village, a National Historic Landmark that interprets the Harmony Society's material culture and industry (Old Economy Village, 2013).

Para, Brazil | The High Line, New York City, 2009

Tabebuia serratifolia and *T. impetiginosa* can reach heights of 50 m and nearly 2 m in diameter, and are among the set of species collectively marketed as ipê lumber. In Brazil, they are known as ipê amarelo (yellow) and ipê roxo (purple) for the brightly coloured flowers that appear in the spring and make them desirable street trees in the country's capitals. Their heartwood is embedded with bio-chemicals known as extractives that make the wood exceptionally rot resistant and their fibre cells have thick walls that render notably hard and dense lumber. This wood is so dense that it sinks in water and is so hard that it must be pre-drilled before assembling. These qualities, derived partially from the very slow growth rate that

FIGURE 2.7 Logs and recently milled ipê (*Tabebuia sp.*) lumber, Belém, Brazil, 2012. Photo by Jane Hutton.

characterizes the species, make the lumber a valuable commodity, worth some $450 US per cubic meter (Lentini et al., 2005, 111) (see Figure 2.7).

The species are distributed from Peru to Mexico, but individuals appear at very low densities – a single mature tree may be found every 3 to 10 ha in the Brazilian Legal Amazon (Schulze et al., 2008, 2077). Like other slow-growing, light-dependent species, populations are composed of large, mature adults and few juveniles (Schulze et al., 2008, 2,081). The demand for valuable and sparsely distributed trees like ipê, and in particular *Swietenia macrophylla* (big leaf mahogany), has been a significant player in illegal logging in the region. The pursuit of high-value species has pushed the logging frontier further and further into unlogged forests, catalysing road construction and agricultural production (Geist and Lambin, 2002, 150). Facing commercial extinction, mahogany was listed in Appendix II of the UN Convention on International Trade in Endangered Species (CITES) in 2002. As ipê species share ecological traits and similar market desirability to mahogany (ipê exports have increased by 500 per cent between 1998 and 2004), ecologist Mark Schulze and his colleagues conclude that ipê 'are the new mahogany' (2008, 2072).

The year 2012 saw the lowest rates of deforestation in the Brazilian Amazon since monitoring began in 1988 (BBC, 2012) due to stricter governmental controls and improved surveillance. Nevertheless, illegal activity is estimated to represent more than 35 per cent of all current logging in the Brazilian Amazon (Lawson and MacFaul, 2010, xvii) and has been linked to a spectrum of exploitative labour practices including slavery by debt (Fearnside, 2008, 30). In the 1990s, campaigns aiming to stem illegal logging and its attendant problems turned from boycotts to market-based strategies linking economic growth with land protection (Zhouri, 2004, 70). Among these was the development of the Forest Stewardship Council (FSC) in 1993, which has become the most recognized third-party forestry certification system in the US. Brazilian certification has been considered a success: 7 million ha of Brazilian forest have been certified by the FSC since the program began (2012, 3).

Ipê and other tropical lumbers rose in popularity in municipal projects in New York City as the widespread use of chromated copper arsenate (CCA) treated softwoods were deemed unsafe. The rot resistance and durability of tropical hardwoods like ipê, cumaru (*Dipteryx odorata*), greenheart (*Ocotea rodiei*) and garapa (*Apuleia leiocarpa*) has made them valuable for public applications. Ipê has become the most popular tropical lumber in the multibillion dollar US decking market (Smith and Cossio, 2008, 21) and has been widely used in prominent, large-scale projects in New York, including the Coney Island boardwalks, site furnishings in the Hudson River Park and decking and benches on the High Line.

The High Line opened in 1934 as part of the West Side Improvement, alleviating the dangerous conflicts between freight and pedestrian traffic of 10th Avenue known as 'Death Avenue'. The 2003 international competition to design a public park on the High Line emblemized the reclamation of the infrastructures of New York's industrial age – infrastructures made obsolete by the shifting of industry elsewhere. The winning

FIGURE 2.8 Rainforests of New York demonstration on Phase 1 of the High Line, New York City, 2009. Rainforest Relief.

team of James Corner Field Operations and Diller Scofidio + Renfro spoke of transforming the elevated rail infrastructure into a 'post-industrial instrument of leisure, life, and growth' (High Line, 2013). A few months after the much celebrated opening of the first phase, members of the Rainforests of New York campaign unfurled a 10-m-long banner on West 17th Street amphitheatre's ipê bleachers that read 'High Crime on the High Line: FSC Lies, Amazon Wood is Not Sustainable' (Figure 2.8).⁶ The campaign, initially supportive of using FSC certified lumber in public works, later became sceptical that certification was a reliable indicator of sustainable production. They cite New York City as the largest consumer of tropical lumber in North America and have been advocating against its use in municipal procurement since 1995 (T. Keating, Rainforest Relief, personal communication, 6 November 2012).

This activism has contributed to a changing public discourse about tropical wood use in New York City. In a 2008 address to the United Nations General Assembly, New York City's Mayor Bloomberg committed to reducing the city's tropical hardwood consumption by 20 per cent (Chan, 2008). New York City Parks and Recreation has since stopped using tropical lumber in park benches and wood alternatives are being tested in pilot projects around the city. Phase 2 of the High Line, launched in 2011, was constructed with oil-treated teak from demolished industrial and agricultural buildings from Indonesia, in lieu of ipê (Associated Press, 2011) (see Figure 2.9).

FIGURE 2.9 Reclaimed teak benches in Phase 2 of the High Line, New York City, 2011. Photo by Jane Hutton.

The High Line, admired for its identity as ultimately local – with its reference to an industrial past and the pioneering vegetation that had occupied the abandoned rail line – was a strategic location from which to invoke the very distant and abstract consequences of ipê use. While many mechanisms such as the FSC work towards minimizing risk of illegally traded wood and promoting best-practice forestry, the worst-case scenario implications – including exacerbation of deforestation and slave labour – are more than daunting issues to navigate.

Conclusion

The landscapes of the Fox Islands, Ambridge and the Amazon Forest of northern Brazil are linked to Central Park, Riverside Park and the High Line, respectively, through the movement of stone, steel beams or lumber from one to the other. They represent a coastal quarry reliant on marine transport, an inland fabricator tied to rail networks and iron suppliers and a distant forest linked by truck and marine networks. Each of the New York City landscapes emblemizes a specific relationship between the city and elsewhere: Central Park brought a civilizing nature to urban dwellers newly divorced from the countryside; the covering of the rail tracks at Riverside Park constructed an automotive escape from the city to exurban areas newly conceived as landscapes of leisure; and the High Line provided a precedent for the re-inhabitation of urban infrastructure made obsolete through the transfer of industrial production overseas. These three cases are linked through design ambitions and infrastructural vision – as Olmsted and Vaux planned both Central and Riverside Parks – and the West Side Improvement encompassed both the renovation of Riverside Park and the construction of the elevated High Line.

Within each set of paired sites, both landscapes are transformed by simultaneous economic forces, but shaped to unequal effects. The concentration of capital and construction in urban projects such as Central Park during the late nineteenth century controlled the cycles of boom and bust employment for granite cutters of the Fox Islands, stimulating the organization of a trade union. The advent of new, cheaper technologies and later access to cheaper labour elsewhere eventually led to the industry's decline in the state of Maine. In the 1930s, New Deal legislation both facilitated the construction of massive infrastructural works as well as stimulated the widespread organization of trade workers. The case of ipê lumber highlights contemporary conditions of globalized neoliberalism with its increasing expansion of material circulation through international markets, the outsourcing of ecological risks and the precarity of workers.

At stake for landscape architecture in considering these expanded material relationships are implications for both theory and practice. Recent contributions within the field have proposed the expansion of sites and scales that landscape architects might practice in, as well as the consideration of ecological processes, material flows and the principles of industrial ecology as generative for design.[7] Conceptualizing the sites of material production as integral – rather than external – to design would shift theoretical concerns of the landscape project without necessarily shifting its site boundary. This has the potential to both examine the ways in which non-adjacent spaces are designed contiguously, but also to speculate about how these reciprocal relationships might be designed themselves.

In practice, while refining material evaluation systems and increasing access to information facilitate knowledgeable timely material specification, a more complex understanding of the social, ecological and economic conditions that form and are formed by materials is needed. As Hartwick argues that a geo-materialist linking of consumers and producers has the potential to stimulate political praxis for the discipline of geography (2000), the same may be said for design. The strategies of the food justice movement are an important precedent in this regard. The movement's insistence and visualization of the linkages between land use, ecological dynamics and social justice has brought the question of where food comes from into daily parlance. The most successful tactics are spatial experiments that challenge normative scales and models of production and propose models of cooperative and sponsored labour. While the construction materials industries are vastly different from agriculture, the very examination of how they work is critical to understanding design today. Greater fluency about the expanded relations of landscape making is necessary to imagine material practice as one that could enact solidarities with workers, other species and landscapes 'elsewhere'.

Acknowledgements

Introductory research for this chapter was first presented at the Landscape and Critical Agency conference organized by Ed Wall, Tim Waterman, Douglas Spencer and Murray Fraser. This chapter was then published in the *Journal of Landscape Architecture* in 2013. I am grateful to the organizers and participants of that conference for their generous and insightful feedback. I would also like to thank Anne Weber, Adrian Blackwell, the editors of *JoLA* and the anonymous reviewer's paper for their constructive comments.

Notes

1 For a comprehensive account of material flow in relation to processes of urbanization, see Baccini and Brunner (2012, 392).
2 The exchange of the commodity in the open market occludes the social relations of production (between worker and consumer), see chapter 1 of Marx (1867).
3 For example, see Mostafavi and Leatherbarrow (1993) and Kirkwood (2004).
4 For example, see Dunnett and Hitchmough (2008) and Margolis and Robinson (2007).
5 For a thorough account of the impacts and issues associated with materials in landscape architecture and evaluation systems, see Calkins (2009). For more information on the Sustainable Sites Initiative, see Calkins (2012).
6 The Rainforests of New York campaign was initiated in 1995 by members of Rainforest Relief and New York Climate Action. See www.rfny.org (accessed 15 January 2013). For a video of the High Line action, see www.youtube.com/watch?v=CsRfrmLW8AU (accessed 15 January 2013).
7 See, for example, Hill (2005), Belanger (2007) and Lister (2006).

References

Associated Press (2011) 'High Line Phase Two Is Open', *Huffington Post*, 7 June, www.huffingtonpost.com/2011/06/07/high-line-phase-two-is-op_n_872651.html (accessed 15 January 2013).

Baccini, P. and Brunner, P. (2012) *Metabolism of the Anthroposphere: Analysis, Evaluation, Design*. Cambridge, MA: MIT Press.

Belanger, P. (2007) 'Landscapes of Disassembly', *Topos* 60, pp. 83–91.

Board of Commissioners of the Central Park (BCCP) (1861) *Fourth Annual Report*. New York: William C. Bryant and Company.

Board of Commissioners of the Central Park (BCCP) (1862) *Fifth Annual Report*. New York: William C. Bryant and Company.

British Broadcasting Corporation (BBC) (2012) 'Amazon Destruction at New Low', 27 November, www.bbc.co.uk/news/world-latin-america-20512722 (accessed 15 January 2013).

Calkins, M. (2009) *Materials for Sustainable Sites: A Complete Guide to the Evaluation, Selection, and Use of Sustainable Construction Materials*. Hoboken, NJ: Wiley.

Calkins, M. (2012) *The Sustainable Sites Handbook*. Hoboken, NJ: Wiley.

Caro, R. (1975) *The Power Broker: Robert Moses and the Fall of New York*. New York: Vintage Books.

Chan, S. (2008) 'Bloomberg Urges U.N. to Act on Climate Change', *New York Times*, 11 February, http://cityroom.blogs.nytimes.com/2008/02/11/bloomberg-urges-un-to-act-on-climate-change/ (accessed 15 January 2013).

Chicago Daily Tribune (1933) 'Industry Spurs Output to Meet Rising Demand', 19 May, p. 30.

Cronon, W. (1991) *Nature's Metropolis: Chicago and the Great West*. New York: W.W. Norton & Company.

Dunnett, N. and Hitchmough, J. (2008) *The Dynamic Landscape: Design, Ecology and Management of Naturalistic Urban Planting*. London: Taylor & Francis.

Fearnside, P. (2008) 'The Roles and Movements of Actors in the Deforestation of Brazilian Amazonia', *Ecology and Society* 13(1), pp. 23–45.

Forestry Stewardship Council (FSC) (2012) 'Global FSC Certificates: Type and Distribution', 27 November, http://ic.fsc.org/facts-figures.19.htm (accessed 15 January 2013).

Gandy, M. (2003) *Concrete and Clay: Reworking Nature in New York*. Cambridge, MA: MIT Press.

Gaynor, P. (2000) 'Something Old Is New Again for American Bridge', *Pittsburgh Post-Gazette*, 23 July, http://old.post-gazette.com/businessnews/20000723ambridge2.asp (accessed 11 February 2013).

Geist, H. and Lambin, E. (2002) 'Proximate Causes and Underlying Driving Forces of Tropical Deforestation', *BioScience* 52(2), pp. 143–150.

Grindle, R. (1976) 'Bodwell Blue: The Story of Vinalhaven's Granite Industry', *Maine Historical Society Quarterly* 16(2), pp. 15–112.

Grindle, R. (1977) 'Tombstones and Paving Blocks: The History of the Maine Granite Industry', *The Courier-Gazette*.

Hartwick, E. (1998) 'Geographies of Consumption: A Commodity-Chain Approach', *Environment and Planning D: Society and Space* 16, pp. 423–437.

Hartwick, E. (2000) 'Towards a Geographical Politics of Consumption', *Environment and Planning A* 32, pp. 1177–1192.

Harvey, D. (1990) 'Between Space and Time: Reflections on the Geographical Imagination', *Annals of the Association of American Geographers* 80(3), pp. 418–434.

Harvey, D. (2006) *Spaces of Global Capitalism: A Theory of Uneven Geographical Development*. London: Verso.

High Line (2013) High Line Team [website], www.thehighline.org/design/design-team-selection/field-operations-diller-scofidio-renfro (accessed 22 February 2013).

Hill, K. (2005) 'Shifting Sites', in C. Burns and A. Kahn (eds) *Site Matters*. New York: Routledge, pp. 131–156.

Historic American Engineering Record (2006) Report on the Henry Hudson Parkway, HAER No-NY 334, http://riverdalenature.org/pdf/Historic-American-Engineering-Record-(HAER)-report-on-the-Henry-Hudson-Parkway.pdf (accessed 10 February 2013).

Hobbs, R. (1981) *Robert Smithson: Sculpture*, 1st edn. Ithaca, NY: Cornell University Press.

Kirkwood, N. (2004) *Weathering and Durability in Landscape Architecture: Fundamentals, Practices, and Case Studies*. Hoboken, NJ: John Wiley.

Latour, B. (2005) *Reassembling the Social: An Introduction to Actor-Network-Theory*. New York: Oxford University Press.

Lawson, S. and MacFaul, L. (2010) *Illegal Logging and Related Trade*. London: Chatham House.

Lentini, M., Pereira, D., Celentano, D. and Pereira, R. (2005) *Fatos Florestais da Amazonia 2005*. Belem: Imazon.

Lister, N. (2006) 'Ecological Design', in R. Cote, J. Tansey and A. Dale (eds) *Linking Industry and Ecology: A Question of Design*. Vancouver: University of British Columbia Press.

Margolis, L. and Robinson, A. (2007) *Living Systems: Innovative Materials and Technologies for Landscape Architecture*. Basel: Birkhauser.

Marx, K. (1867/2011) *Capital: Critique of Political Economy*. New York: Dover Publications.

Merrill, G.P. (1891) *Stones for Building and Decoration*. New York: John Wiley & Sons.

Mostafavi, M. and Leatherbarrow, D. (1993) *On Weathering: The Life of Buildings in Time*. Cambridge, MA: MIT Press.

New York Times (1934) 'West Side Project Ready to Hire 4,000: Needy to Get the Jobs', 28 May.

New York Times (1936a) 'Work Speeded on City's Riverside Park', 19 January, p. E10.

New York Times (1936b) 'Extending West Side Highway to the North', 5 January, p. xx6.

Old Economy Village (2013) www.oldeconomyvillage.org/ (accessed 11 February 2013).

Rich, G. (1892) 'The Granite Industry of New England', *New England Magazine*, February.

Schulze, M., Grogan, J., Uhl, C., Lentini, M. and Vidal, E. (2008) 'Evaluating Ipê (*Tabebuia bignoniaceae*) Logging in Amazonia: Sustainable Management or Catalyst for Forest Degradation?', *Biological Conservation* 141(8), pp. 2071–2085.

Slater, L. (2008) *Ambridge: Images of America: Pennsylvania*. Charleston, SC: Arcadia Publishing.

Smith, B. and Cossio, V. (2008) 'Competitiveness of Forest Products at Global Markets', Draft report, Food and Agriculture Organization of the United Nations, http://foris.fao.org/preview/18280-0a834cc976bb31035ac3eeb948a1839ed.pdf (accessed 15 January 2013).

Sweeny, H. (1937) *West Side Improvement: Published on the Occasion of the Opening October 12th, 1937*. New York: Moore Press.

Vogel, J. (1932) 'Opening Ways from the City to North, East, South, West', *New York Times*, 10 January, p. A24.

Wall Street Journal (1933) 'Four More Steel Plants Shut Down: Man Shot, Three Beaten in Clash at Ambridge, Pa., Wednesday – Spang Chalfant Picketed', 5 October.

Wall Street Journal (1936) 'Structural Steel Orders', 22 September, p. 3.

Warren, K. (1973) *The American Steel Industry 1850–1970: A Geographical Interpretation*. Oxford: Clarendon Press.

Wilkinson, B. (2005) 'Humans as Geologic Agents: A Deep-Time Perspective', *Geology* 33(3), pp. 161–164.

Zhouri, A. (2004) 'Global–Local Amazon Politics: Conflicting Paradigms in the Rainforest Campaign', *Theory, Culture & Society* 21(2), pp. 69–89.

3
AGENCY, ADVOCACY, VOCABULARY
Three landscape projects

Jane Wolff

When New Orleans flooded after Hurricane Katrina, and when New York flooded after Superstorm Sandy, and when Toronto flooded after a summer storm that wasn't even a hurricane,[1] people asked the same question over dinner and in the media: *how could this have happened?* But each of these calamities might have provoked – should have provoked – a different question, one that was more apt: *why were we surprised?*

We were surprised because most of us don't have the background to make sense of the complicated places we live in. In New Orleans and New York and Toronto and in many other places across North America, a disaster that caught the public unaware was actually the predictable outcome of reciprocal interactions between a dynamic landscape and the steps people took in order to live there in large numbers. These cities, like most contemporary landscapes, are ecological hybrids. Neither nature nor culture, their inseparable mix of geographical circumstances, social demands and environmental processes is hard to unravel and hard to describe. Though landscape designers, planners, engineers, scientists and scholars all have ways of representing and understanding ecological hybridity,[2] the language of their discussions is usually too specialized for broad audiences. For people who aren't experts, hybrid landscapes are almost always indecipherable puzzles. This obscurity means that again and again, cities and regions are thrown into disarray and citizens are shocked and confused by events that might have been anticipated, prepared for, mitigated or even prevented.

The language gap between experts and citizens limits the effective discussion and management of hybrid landscapes. Technically sophisticated representations tend to be inaccessible to outsiders, and citizens rarely have opportunities to share their experiential knowledge of a place with experts. All too often, laymen aren't able to grasp experts' insights fully, experts aren't able to offer what they know for general debate, and valuable observations made in the course of ordinary life are left out of the conversation entirely. As a result, no one has all the information necessary to engage in nuanced conversation about immediate choices and long-term possibilities.

This is a political problem. Hybrid landscapes are the products of collective, incremental (and often invisible) decisions by individuals, markets and governments; their intense use means plural constituencies whose interests in and understandings of the same environment often vary widely; and their dilemmas will only be made more pressing by sea-level rise and climate change. A more resilient future for these places depends both on better choices at the grassroots level and on widespread support for difficult policy changes. Inclusive, informed public discussion about what's next is an urgent necessity. Translating between the technical, conceptual knowledge of experts and the experiential knowledge of citizens offers the possibility to extend agency on both sides.[3]

The need for translation raises a compelling opportunity for landscape architecture, where ecological hybridity is a given from the scale of the body – say, in a row of pleached plane trees – to the scale of the region – say, in the planning of an inhabited river basin. If landscape architects turn their observational and representational skills[4] to the development of shared language for difficult places, they have the chance to extend the agency of design. Articulating the terms of public debate about large-scale landscapes is a way for the discipline to engage problems and arenas that have traditionally been beyond its reach. Whatever our roles in the evolution of landscapes, as experts or as citizens, our means and modes of acting depend on what we argue for, and our arguments for possibilities in the future arise, one way or another, from the words and images we use to describe the places we know now. Agency arises from advocacy, and advocacy depends on vocabulary.

Three case studies in landscape vocabulary

The idea of shared language is straightforward, but the development of visual and verbal terms (and the tools to disseminate them) is a complicated endeavour. It demands the thorough observation of situations with many different qualities and meanings, the evaluation of their layers of significance and the distillation of what's essential about them. It also requires the development of integrated visual and verbal representations that are both simple enough to be understood by people new to the subject matter and sophisticated enough to convey technical information. It depends on flexible organizing schemes that allow content to be combined and recombined according to the expertise, interests, points of view and ambitions of different readers. Hybrid landscapes are compelling because they're nuanced, and their meaning can't be brought to public discussion effectively without multidimensional tools.

This chapter brings together three case studies from an ongoing, long-term investigation in the development of vocabulary for hybrid landscapes. Each case deals with a landscape that's complicated, contested and subject to pressure from sea-level rise and climate change.[5] Each represents a vivid, high-stakes version of conditions that can be found in many other places. *Gutter to Gulf* (Wolff et al., 2011) discusses hydrological and hydraulic systems in a delta city: post-Katrina New Orleans. *Bay Lexicon* (Wolff, 2013) documents an urban waterfront that forms the centre of a metropolis: the edge of San Francisco Bay. *Delta Primer* (Wolff, 2003) addresses a rural landscape subject to conflicts between local claims and vast resource needs from far away: the California Delta. The future of each of these places depends on some combination of political will, technical possibility and government policy, but their trajectories are different, and each landscape asks a different question about the definition, design and dissemination of useful language.

Because agency, advocacy and vocabulary depend on context, the projects differ in their scopes, media, venues and audiences. They share the same aim, the articulation of nuanced, place-based vocabulary that makes rigorous technical information available to a wide range of readers. They use the same basic devices – texts and drawings that borrow from representational conventions familiar to many people – and though they depend on different strategies and methods, each proposes a type of visual and verbal language that's both simple and sophisticated. Together, they suggest a consistent, broadly applicable set of questions, strategies and methods for documenting, describing and defining hybrid ecologies in ways that are accessible to people with different points of view, different sorts and levels of expertise and different stakes in the future.

Gutter to Gulf: *vocabulary for a delta city*

Gutter to Gulf is a website that provides previously unavailable information about the hydrology and hydraulics of New Orleans. The project was a response to a significant gap in the public understanding of the city's landscape systems: by the time Hurricane Katrina hit, the collective memory that New Orleans belongs to the regional ecology of the Mississippi Delta was long gone.

Until the early twentieth century, urban development in New Orleans was largely confined to high ground along the banks of the Mississippi River and one of its abandoned ancient channels. The large-scale reclamation of the city's back-of-town swamps began in the early twentieth century, and by the middle 1950s, central New Orleans was fully built out. The removal of water from that swampy ground produced subsidence, and after a hundred years of mechanical drainage, the city is shaped like a bowl, protected by levees and floodwalls from the river, canals and lake that form its boundaries.[6] All of the rain that falls within these constructed edges must be evacuated from the city by pumping, and the city's pumps, pipes and channels are not large or powerful enough to remove or store even ordinary amounts of precipitation. Low-level flooding is a constant, everyday threat. This state of affairs emerged gradually and incrementally enough to be taken for granted, and faith in technological solutions led to the disavowal of risk. Katrina served as a painful reminder of the system's vulnerability. The most significant mechanism of flooding in central New Orleans was intrinsic to the mechanics of reclamation: a storm surge from the Gulf of Mexico backed up into two of the city's main drainage canals, breaching their walls, and overwhelmed the pumps that usually keep the city more or less dry.

After the hurricane, official planning and recovery efforts concentrated on replacing and restoring buildings. When I went to work with grassroots organizations in New Orleans eighteen months after the storm,[7] the subject of urban water management was not on the table, and advocates for the rehabilitation of the metropolitan landscape were hard put to find even basic information about infrastructure and ecology. The city's systems were so opaque – and its physical geography so strange – that it was impossible to trace the course of a raindrop from the gutter where it fell to its end in the Gulf of Mexico. My colleagues Elise Shelley, Derek Hoeferlin and I saw this information gap as a political problem, an impediment to responsible design proposals and an opportunity for design research. Working with our students,[8] we set out to uncover, document and explain the city's hydrological and hydraulic systems in terms that were both technically rigorous and clear enough for ordinary citizens to understand. Our goal was to make information essential to the city's rehabilitation available and accessible to anyone who wanted it: citizens, designers, planners, policy-makers and politicians. We saw this endeavour as an essential step toward the thoughtful discussion of New Orleans's relationship to water both in technical and policy circles and in broader public debate.

Working from a diagram obtained from New Orleans's Sewerage and Water Board, primary and secondary source research on the history, technology and present state of the storm sewer system, city policy documents and field observation, our students developed drawings and models that explained the role of water in the city's landscape at multiple scales, from individual sites to the city's main internal drainage basin. The students also carried out research on and made drawings to explain regional and continental phenomena that affect conditions in New Orleans.

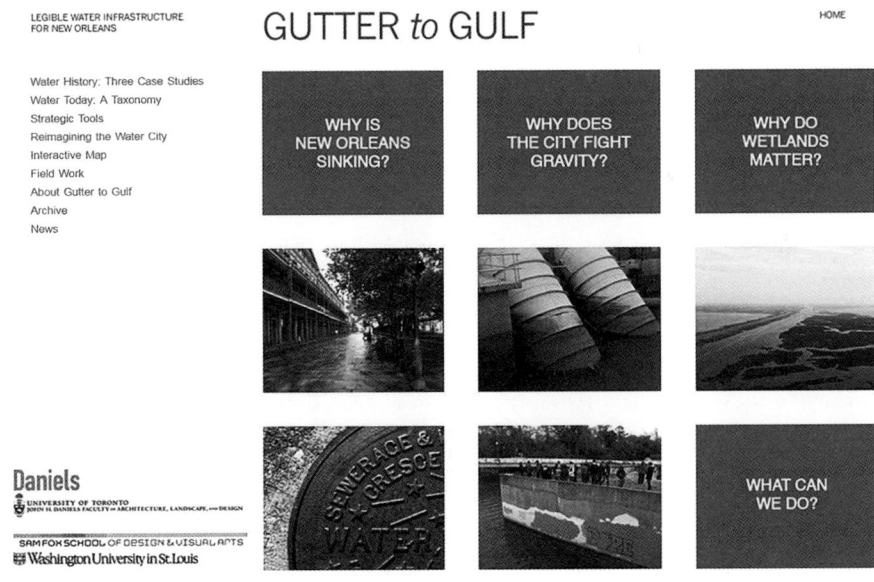

FIGURE 3.1 Home page, *Gutter to Gulf* website. The main body of the site's home page links to text, drawings, animations and photographs that address basic questions about water systems in New Orleans. An additional menu connects visitors to historical case studies; a taxonomy of water infrastructure; digital and physical tools for discussing alternative strategies; design proposals at scales from the individual lot to the city and region; and downloadable field guides to the city's hydraulic service areas.

The flood threat created locally by subsidence has been exacerbated by long-term regional and continental landscape management strategies. The southern Louisiana wetlands that protect the city from storm surges in the Gulf of Mexico have been decimated by the closing of the Mississippi's distributary channels, or bayous; the dredging of canals for navigation; the construction of oil and gas pipelines; and saltwater intrusion. The development of the Mississippi's drainage basin has put more water in the river, even at ordinary stages and the constriction of the river's channel by levees creates higher and higher water levels during floods.

Ms Shelley, Mr Hoeferlin and I curated, composed and wrote about this work in the website *Gutter to Gulf*. The site is organized thematically around four questions of importance to anyone with a stake in the city's future. Three address water-driven dilemmas: 'Why is New Orleans sinking?'; 'Why does the city fight gravity?'; and 'Why do wetlands matter?' The fourth, 'What can we do?', advocates for a water management plan to guide New Orleans's rehabilitation and redevelopment.

The site uses text, drawings, animated diagrams and interactive maps to explain the present circumstances of the city's hydraulic and hydrological systems; those systems' historical evolution; and their relation to institutions and districts for policy, management and governance.

The site also includes detailed field guides for people who want to explore the city's drainage system for themselves. Organized according to New Orleans's

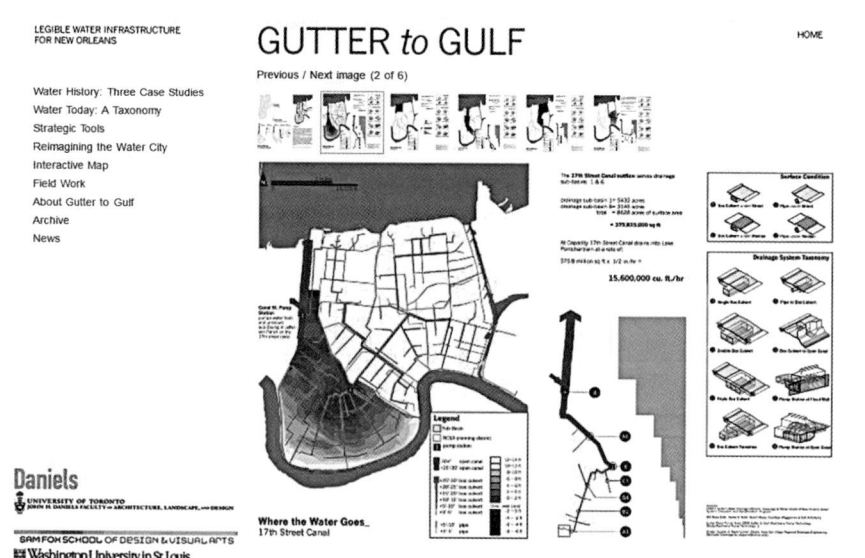

FIGURE 3.2   Water taxonomy: subsurface drainage system. The map of central New Orleans is redrawn and annotated to reveal the structure of the city's drainage units, including subsurface drainage routes, pumps, outlet canals and pipe and vault types and sizes. It also describes the volume of water moving through the system and the elevation of the ground surface. Drawings by Justin Cheung, Marc Hardiejowski, Juan Robles and Scott Rosin.

hydraulic subdivisions, or drainage service areas, the guides offer comparable analytical data about landscape characteristics such as surface area, drainage capacity and elevation range; explain the city's evolution in terms of urban, landscape and architectural form; describe drainage infrastructure and policy structures in detail; and provide routes among and descriptions of three sites essential to each drainage service area. Looking forward, *Gutter to Gulf* offers a collection of landscape, infrastructure and urban design proposals that demonstrate strategies for ameliorating the city's flooding problems at everyday and extreme levels.

New Orleans and southern Louisiana are landscapes of national significance economically, ecologically and culturally, and *Gutter to Gulf* advocates for public education, synthetic planning and political action toward a more resilient future there. Disseminated to experts through workshops on water and design in the city in 2009 and 2010,[9] the project was used as technical and scholarly base information and education and outreach material for the *Greater New Orleans Urban Water Plan* (State of Louisiana Office of Community Development – Disaster Recovery, 2013). It has become the first-line reference for activists and professionals working on water-related issues in the city, and it has been used as a case study by academics and students all over the world. Widely hailed as a valuable tool by grassroots organizations, engineers, designers, geographers and city planning officials for its precise, clear explanations of circumstances that had been opaque, the project mobilized design research in academic

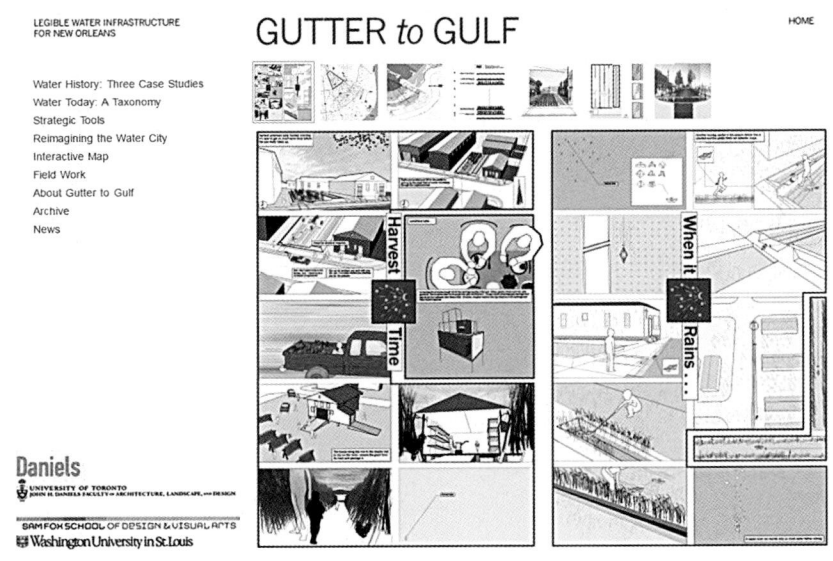

FIGURE 3.3 A proposal for reimagining the urban water landscape. This rice farm proposes small-scale cooperative agriculture to address problems of water storage and economic opportunity. Surface channels transport storm water to rice paddies cultivated on vacant properties. The cultivation cycle is calibrated to seasonal rainfall patterns, and even small areas can generate profit without intense labour. Project by Adam Bobbette and Karen May.

institutions as a tool for critical agency: a labour pool of professors and students was able to provide essential information for public and professional debate about the future of a difficult, important place.

Bay Lexicon: *vocabulary for a metropolitan waterfront*

Bay Lexicon is a visual dictionary of the edge between San Francisco and San Francisco Bay. Developed as an exhibit at the Exploratorium, an innovative science museum in San Francisco,[10] the project articulates vocabulary that helps people recognize the ambiguous, complex, varied ecology of a landscape that's loved better than it's understood.

San Francisco and the bay are so powerful as scenery that their natural and cultural dynamics frequently escape notice. The bay and its edges comprise one of California's most productive ecological habitats. Defined by the low mountains of the Coast Ranges, the bay landscape is also part of the largest delta and estuary system on the west coast of the Americas: it provides the only connection between California's Great Central Valley and the Pacific Ocean. It has had national military significance since the European colonization of North America's west coast, and the US Army and Navy have left an environmental legacy that is positive in some ways and problematic in others. Because it links the state's resource-rich interior to national and international waters, it has been an economic engine for metropolitan development, and it has been transformed by navigation, speculation, commerce and industry. Now the centre of a nine-county metropolitan region, the bay and its edges are subject to intense use and competing demands. Pressure for change from sea-level rise, urbanization, economic booms (and busts), environmental politics and public regulation will only increase, and the situation is more complicated than it looks.

Bay Lexicon arose from an extended conversation with Susan Schwartzenberg and Peter Richards, curators at the Exploratorium. Frank Oppenheimer, a former Manhattan Project physicist, founded the museum in 1969 because he believed that widespread scientific literacy was politically essential in an era driven by technology. Ms Schwartzenberg and Mr Richards wanted to expand the institution's traditional focus on the basic sciences into environmental studies. They undertook their work as climate change was entering widespread consciousness, and their aim was to foster stewardship of the bay by educating citizens about its dynamics.

Building on Oppenheimer's commitment to direct perception as an essential means of learning, *Bay Lexicon* began with the premise that language is the first tool for perception: we cannot recognize what we cannot name. The project comprises a set of illustrated flash cards that examine tangible artefacts and phenomena and connect them to the complex and often invisible practices, processes and relationships transforming San Francisco Bay. The flash cards all begin with a scene that's visible either from the windows of the Exploratorium's Bay Observatory gallery or along the shoreline.

Each card defines language for the bay at different levels of complexity and using different means. The first is visual: many of the drawings reveal essential structures

Agency, advocacy, vocabulary **41**

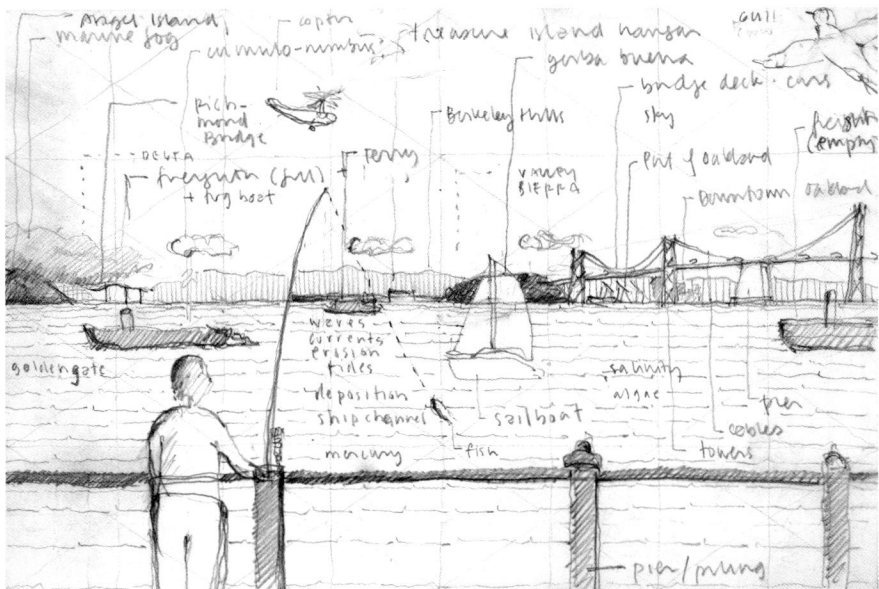

FIGURE 3.4 Sketch of vocabulary subjects, *Bay Lexicon*. *Bay Lexicon* distils and examines inventories of objects and phenomena an observer can see along the shoreline. To define a working vocabulary for the bay's edge, each flash card synthesizes research about the forces and processes that underlie what's visible.

and phenomena that are hidden from view, like seawalls and constructed land, or that can only be perceived over time, like changes in the tide. The second relates text and image: the components of each drawing are identified, named and described. The third uses text to examine image: an open question about each drawing's subject is linked to a brief meditation on its possible meanings. The fourth is taxonomic: each flash card is connected to three keywords related to the bay landscape, and together, all of the flash cards grouped under that keyword provide an operational definition of its meaning in the particular context of San Francisco Bay and its shoreline.

The flash cards reveal layers of complexity beneath and behind what seem to be straightforward situations and words. Even the terms 'land' and 'water' turn out to be ambiguous. The edge of the bay, so easily identified as a line on most maps of San Francisco, is actually a fluctuating zone whose current condition emerged gradually from the combined forces of property law, greed, convenience and landfill. Water permeates the apparently dry ground of the city, flowing through pipes, soil and cracks in walls, while land, dispersed as sand and sediment and carried by the energy of rivers and tides, travels into, around and out of the bay. The flash cards demonstrate ways in which a look out the gallery window or a walk along the shore brings a reader into contact with a tangle of cultural and environmental processes. For instance, turbidity that makes shadows on the water comes in large part from nineteenth-century hydraulic mining in the Sierra Nevada; longshore drift produces a new wildlife habitat on a derelict bridge abutment; geomorphology set

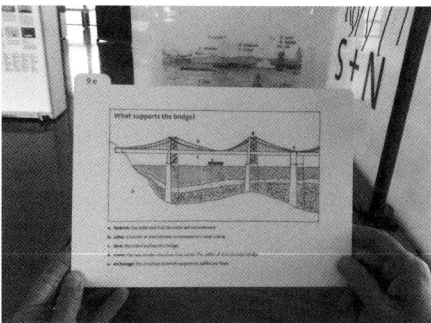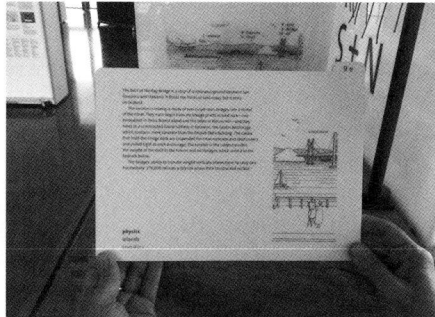

FIGURE 3.5   A flash card's front and back. The front and back of each flash card offer complementary definitions of landscape terms. For example, this card's front includes an annotated drawing showing how the Bay Bridge is supported not only by its towers and cables but also by the bedrock beneath its piers. The back explains the system and its capacities verbally, locates the bridge in a view of the bay and identifies three keywords that classify the definition in relation to other flash cards.

up the conditions for a financial, commercial and military nexus of international importance; globalization turned the waterfront from a landscape of production into one of consumption. Each of the flash cards discusses the landscape's hybrid ecology at a range of scales and in the contexts of space and time, and each card identifies ordinary people as agents in the landscape's continuous evolution.

Launched in 2014, *Bay Lexicon* serves as the physical and conceptual centre of the Exploratorium's Bay Observatory gallery, which includes a range of resources for interpreting the bay landscape. In addition to engaging visitors to the Observatory, the exhibit serves as a teaching tool for staff members and science instructors associated with the museum. The gallery's programme has expanded beyond the display of exhibits to include lecture series and public discussions about the future of San Francisco Bay, particularly in relation to climate change, and *Bay Lexicon* has been a point of reference for those events. At present, the content of the exhibit is being extended into a book manuscript; when it is published, the project will serve as a field guide to the hybrid ecology of the bay landscape.

Delta Primer: *vocabulary for a resource landscape*

Delta Primer is a book and deck of cards that describe the complex, fragile landscape of the California Delta, north and east of San Francisco. The ecological and economic lynchpin of California, the Delta is fiercely contested, and the project was designed to address two dilemmas going forward. First, the vocabularies of the interest groups directly involved in negotiating the landscape's trajectory were so different that compromise seemed impossible. Second, the great majority of citizens who would be asked to vote on the region's future didn't even know that the Delta existed.

Agency, advocacy, vocabulary 43

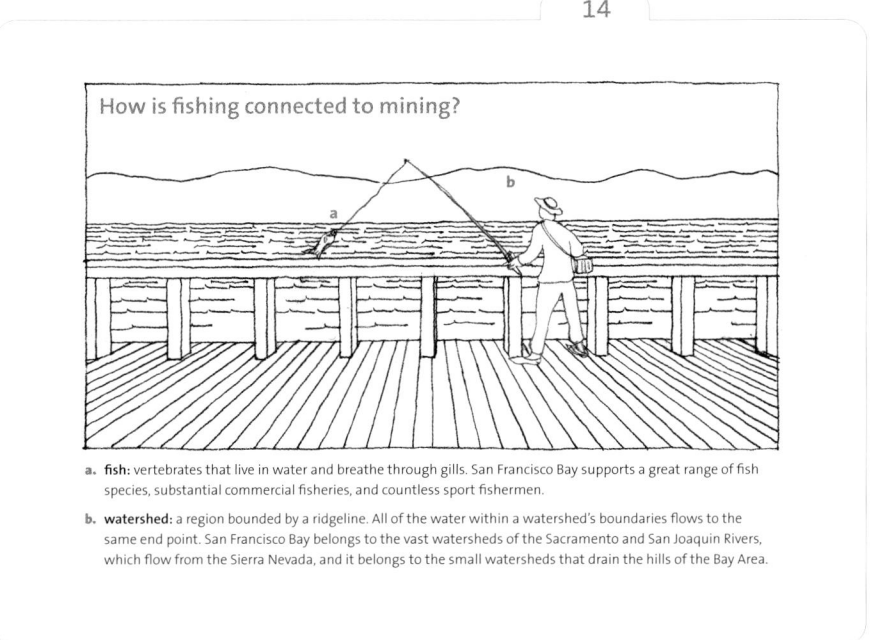

FIGURE 3.6 Flash card: fishing. This flash card speaks to the ways in which human and non-human processes interact over significant distances and long periods of time. It explains that in the nineteenth century mercury mined at San Jose was used to catalyse gold in the Sierra Nevada and then travelled into San Francisco Bay. Absorbed by microscopic biota, it entered the food chain and persists in the fish that people catch today.

The California Delta, where the Sacramento and San Joaquin Rivers meet San Francisco Bay, is one of the most complicated and intensely disputed landscapes in the United States. The largest tidal estuary on the west coast of North America, it has subsided to as much as 20 feet below sea level because of agricultural reclamation. Protected by a tenuous system of levees, it remains extremely productive farmland. In the last sixty years it has gained value in other ways: as a site for northern California's explosive urban growth, a recreational resource, a commercial shipping route, a habitat for protected fish species and the centrepiece of the system that supplies water to the farms and cities of southern California. The landscape's resources are limited, and the demands of its constituencies are often in conflict. In the late 1990s, when I got to know the California Delta, it was apparent that official planning processes were being stymied not only by contradictory goals for the landscape but also by the language of interest groups. Each of the Delta's constituencies tended to describe the landscape in terms of its own functional requirements, and their wide-ranging vocabularies offered no common ground for discussion. Beyond that, although the region's future would depend on public support for ballot initiatives, the planning process was not intelligible to most of the state's voters, and the landscape remained largely unknown to the 23 million people who depended on it directly. The potential costs of this invisibility were high. What happened in the Delta would have major long-term consequences for all of California.

I conceived *Delta Primer* as a political tool: its aims are to make the region's dilemmas visible to diverse audiences and to transcend the usual boundaries of interest groups. The project divides a map of the region into the same number of

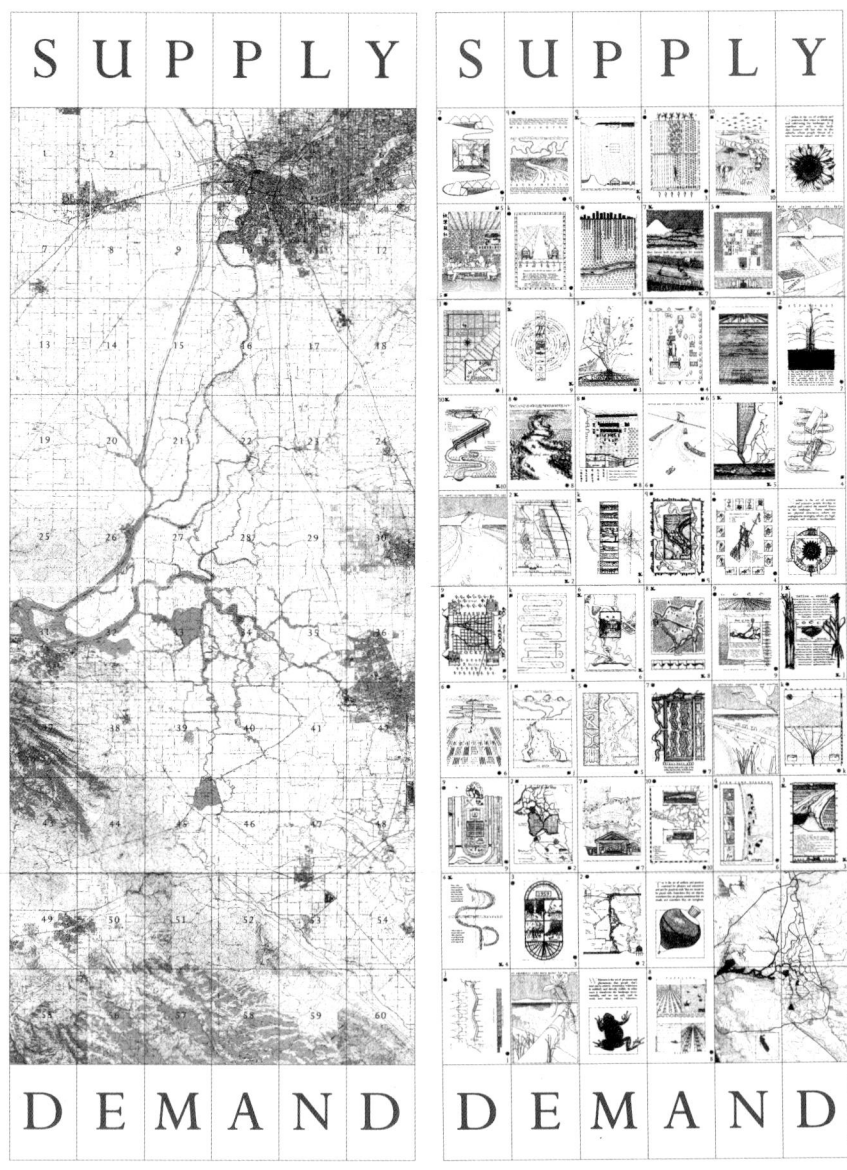

FIGURE 3.7 Playing card map: front and back. The Delta's water supply comes from the north of California, demand comes from southern California.

pieces as a deck of playing cards. Each piece of the map is paired with a documentary drawing that describes a landscape artefact, practice or process located there; together, the drawings comprise a vocabulary for the Delta.

This vocabulary is organized into four categories that describe the landscape and its constituencies. In the official planning process, those descriptions were made only in terms of function and interest groups, and they were conflicting and divisive. Instead, *Delta Primer* proposed four descriptions that had emerged from the documentary stories behind its drawings. The first is that the landscape constitutes a garden people can cultivate and inhabit, someplace that sustains them; the second, that it is a machine people can engineer and transform for any purpose; the third, that it is a wilderness people will never fully control; and the fourth, that it is a toy, someplace where people amuse themselves and find pleasure.

These ideas are synthetic and open to interpretation, and anyone with a stake in the region can see his or her interests represented in every category. Together, the drawings make up a standard deck of playing cards. Each of the big ideas about the Delta – that it is a garden, a machine, a wilderness and a toy – is a suit in the deck. Each drawing is assigned to a suit, and within suits, drawings are ranked according to scale. Any hand in this deck offers a combination of places, processes, practices and artefacts that might exist together. Any player has the ability to assemble a vision of what might be. Every game is a negotiation in which the choice of one card is made at the expense of another and in which there are different ways to arrive at solutions of equal value.

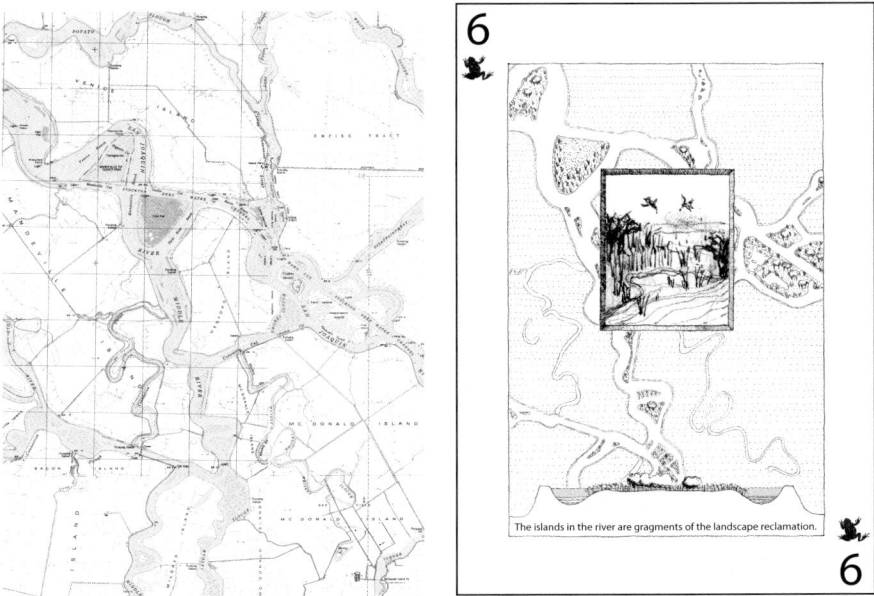

FIGURE 3.8 Playing card: 6 of Wilderness. This playing card from the Wilderness suit describes the small islands in the Delta's river channels. Too small for profitable reclamation, they were abandoned to nature and remain as fragments of the landscape before its agricultural transformation.

FIGURE 3.9 A straight hand from the *Delta Primer* deck. The *Delta Primer* deck can be used to play any standard game of cards. Each hand assembles a different combination of possibilities for the future.

Delta Primer was published as a book and deck of cards at the end of 2003. It attracted attention from a wide range of audiences: policy-makers, journalists, environmentalists, civil servants, engineers, scientists, historians, artists, designers, teachers and children. The project did not solve the region's problems, but it found a place in public discussions of the future, and it led to invitations to participate in conferences and conversations that did not ordinarily include designers. People involved at high levels in the negotiation of the Delta's future, including the executive director of the consortium of state and federal agencies that manage the region, the executive director of the Delta Protection Commission, the head of the Bay Institute's Delta programs and the manager of the Nature Conservancy's efforts in the region, told me that the project had changed their points of view. Twelve years after the project's publication, the Delta is more threatened than ever. *Delta Primer* remains the most thorough document of its vanishing cultural landscape.[11]

In search of means and methods

The distillation of consistent methods from varied examples is a tricky business. Making progress toward ecological integrity in hybrid landscapes means coming to terms with their specific qualities and characteristics. Places like the California Delta and New

Orleans are in trouble today largely because modernist, technocratic attitudes tried to make landscapes uniform at the expense of their basic ecological characters. Nuanced vocabulary for hybrid landscapes can't – and shouldn't – be standardized, but the work to produce *Gutter to Gulf, Bay Lexicon* and *Delta Primer* has suggested some useful, adaptable, broadly applicable strategies for observing, analysing and representing both the typical tendencies and particular idiosyncrasies of hybrid ecological systems.

Each project began with a question about the landscape that seemed simple but wasn't fully answered by standard, commonly available representations. For *Gutter to Gulf*, the question was 'Where does water go?' For *Bay Lexicon*, it was 'Where is the edge of the bay?' For *Delta Primer*, it was 'Whose landscape is this, and why?' In each case, the question led to the reconsideration of an ordinary map as a spatial index for vocabulary that described the landscape as a compendium of artefacts, conditions, processes and phenomena and that could be used to tell a range of meaningful stories about the place.

The definition of a vocabulary for each landscape emerged from systematic enquiry about its material and experiential elements. This meant asking a series of questions:

> What are the components of the system, and how are they related spatially?
>
> Where and how can they be seen, perceived, observed and experienced?
>
> How are they related by dynamic processes at scales from the very small to the very large?
>
> How did they come to be, and how are they understood in the context of time?
>
> In what ways are they dynamic?
>
> To what extent are they predictable?
>
> How might they exist in different relationships to each other?
>
> Who inhabits the system?
>
> Who has had the power to change it?

The next step was to pursue these questions through a range of methods: field observation, interviews, the review of scholarly documents and engagement with popular representations about each place. In each case, a place-based vocabulary emerged from the research, and in each case, this vocabulary could be organized and mobilized to tell meaningful stories about the landscape's past and present circumstances and to speculate on its possible futures.

Each of these projects distils analytical fieldwork about topics many people find difficult to understand into tools that make their subjects widely accessible. These tools aren't hard to use, but they require some time and attention: the dilemmas presented by hybrid landscapes don't lend themselves to soundbite explanations or instant answers. Any of the projects can be understood as a shortcut that offers essential information to people who wouldn't otherwise have the opportunity to encounter it in a manageable form. All of the projects represent these stories through text and images that are both technically sophisticated and easily intelligible. They all use simple drawings and writing styles that borrow from graphic and textual conventions most people understand. They all announce themselves as interpretations, and they do not attempt to present themselves as documents of an inarguable truth. They all discuss information at a range of scales from the small to the vast. They all situate particular details and conditions in the context of more general information. They are all inflected by local cultures. Most important, they all have taxonomic systems that permit alternatives for the organization and reorganization of their content. This conceptual and physical flexibility allows plural and varied interpretation of what exists. It can express and convey different messages, and it permits the constellation of varied ideas about the future. These tools are not only intended for the analysis of current circumstances but also to enable the 'work of evaluation that lies between is and ought' (Sayer, 2011, 16). Their goal is not to predetermine choice but to articulate questions and possibilities that can help people decide for themselves what matters. Different people find different meaning and value in the same landscapes, and every complex decision involves the negotiation between reasoned absolutes and particular contingencies.[12] That range must be represented in order to enable public discussion that is both inclusive and ethical.

Looking forward

In the early days of landscape architecture in North America, Frederick Law Olmsted and Charles Eliot worked not only as designers but also as advocates for public awareness, public value and public protection of landscapes they considered essential to their culture.[13] They claimed a large territory for the discipline, one that engaged the collective and political arenas in which questions of social importance were debated and decided. The translation work in *Gutter to Gulf*, *Bay Lexicon* and *Delta Primer* aims to carry those claims forward. Just as large-scale public health studies complement the work of individual doctors with individual patients, the development of broadly shared environmental language offers landscape architecture the possibility of agency at scales beyond individual sites and immediate problems. The landscapes most people know are shaped by forces outside design – politics, economics, the cultural status of land as a commodity and the deeply entrenched mythology of endless resources. Expanding landscape architecture's specific tasks from designing sites to designing conversations means giving up the control of physical form,[14] but it gives the discipline the chance to influence landscapes that have traditionally been beyond its reach.[15]

The three case studies in this chapter were undertaken to address an urgent task for designers: engaging broad audiences and diverse constituencies in the discussion of complex, contested landscapes. Conceived and executed outside any official systems or institutions that manage landscapes, they were designed to test unconventional methods of observation, analysis and documentation. Their strategies vary, but their ends are the same: to characterize and redefine a place in terms of its essential qualities, to imagine it in a different way, to stop taking it for granted.

Together, the projects have demonstrated ad hoc that synthetic, place-based language has the ability to speak to different audiences, and they suggest ways in which the open-ended, critical examination of places we already know offers clues about what they might become. The next challenge is to find a place for this kind of work inside the policy system, where generic analytical language and methods tend to smooth away particularities and make places more and more alike.

Engaging policy does not mean abandoning the public. Large-scale change in hybrid landscapes will depend on a combination of regulation by governments and aggregated decisions by citizens, and vocabulary that can be shared by the grassroots and the power structure is the first step toward bringing those forces together. And working toward resilience does not only mean creating places that can survive and recover from environmental catastrophe. Resilience also means building a culture whose understanding – whose vocabulary – can adapt to and flourish in the complicated, idiosyncratic, hybrid landscapes it inhabits.

Notes

1 These events occurred in August 2005, October 2012 and July 2013, respectively.
2 The principle that landscapes are shaped by both natural and cultural forces has been more and more widely accepted in the last fifteen years. It emerged over several decades in a range of disciplines, and though the literature is too extensive to summarize here, I wanted to mention a few landmarks. In landscape architecture and planning, Anne Spirn's *Granite Garden* (1985) and Michael Hough's *City Form and Natural Processes* (2004) both articulated the notion of the city as a hybrid. In environmental history, William Cronon's *Nature's Metropolis* (1991) discussed the interactive relationships between human and non-human forces in the growth of Chicago. In ecology, systems ecologists C.S. Holling and Eugene Odum (Goldberg and Holling, 1971; Odum, 1977) examined the role of anthropogenic forces in ecosystems. In philosophy, Felix Guattari (2008) proposed that the environment is shaped by human and non-human forces.
3 Jeremy Till (2009) has written extensively about the value of including the knowledge of citizen experts in design processes.
4 Landscape architects are trained to observe complex ecological conditions and to document them through drawing and modelmaking. Many landscape architectural documents are too technical for lay audiences, but the skills involved in their production could be used to make representations that are much more widely accessible.
5 These case studies all concern landscapes in the United States, where weak regulation and the absence of a clear tradition of planning make landscape literacy among the general public an urgent political and ecological problem.
6 Craig E. Colton's *An Unnatural Metropolis: Wresting New Orleans from Nature* (2005) and Richard Campanella's *Time and Place in New Orleans: Past Geographies in the Present Day* (2002) provide detailed accounts of the development of the city's drainage infrastructure. Campanella's work, which also includes *Bienville's Dilemma: A Historical Geography of New Orleans* (2008), *Geographies of New Orleans: Urban Fabrics Before the Storm* (2006) and *New Orleans Then and Now* (with Marina Campanella, 1999) comprises an encyclopaedic account of New Orleans's evolution. For a discussion of the city's recent dilemmas with respect to ecology and design, please see my essay 'Cultural Landscapes and Dynamic Ecologies: Lessons from New Orleans' (Wolff, 2014), in Chris Reed and Nina Marie Lister's *Projective Ecologies* (2014).
7 Between 2007 and 2011, I worked with several grassroots organizations involved in the rehabilitation of New Orleans's urban landscape. My participation began with an invitation by Longue Vue House and Gardens to lead their joint effort with the Pontilly Disaster Collaborative to develop rehabilitation strategies for the neighbourhoods of Pontchartrain

Park and Gentilly Woods. This endeavour (which expanded to include Julie Bargmann, Elizabeth Meyer and William Morrish and their students at the University of Virginia, Elizabeth Mossop and her students at Louisiana State University and Mia Lehrer, principal of Mia Lehrer+Associates) was documented in the *Ponchartrain Park + Gentilly Woods Landscape Manual* (Wolff and Reese, 2009). From 2008 to 2010, I was part of the Dutch Dialogues, a series of workshops that enabled discussion between North American and Dutch experts on water management and urban design in New Orleans. In 2009, I worked with Austin Allen, Walter Hood and Elizabeth Mossop on a landscape proposal to the Make It Right Foundation for the renewal of the Lower Ninth Ward. In 2011, I served as a technical advisor to the consultant and advocacy team that instigated and prepared the New Orleans Water Plan.

8 This work was carried out through a series of studio courses in landscape architecture at the University of Toronto and in architecture and urban design at Washington University in St Louis between 2009 and 2013.
9 The initiative provided base information for the Dutch Dialogues workshops sponsored by the Kingdom of the Netherlands, the American Planning Association and Waggoner and Ball Architects.
10 This work was commissioned as material for a permanent exhibition on landscape observation and forms the basis of a book manuscript in process.
11 The book was the first comprehensive scholarly study of the region's cultural landscape since 1957, when John Thompson completed his doctoral thesis 'The Settlement Geography of the Sacramento-San Joaquin Delta, California' at Stanford.
12 In his book *Architecture Depends*, Jeremy Till (2009) discusses the difficult relationship between strong knowledge of absolutes and the particular circumstances from any given problem.
13 Olmsted's status as a public intellectual helped him to create a broad understanding of the need for designed public landscapes in the United States. In his book *Walks and Talks of an American Farmer in England* (1852), he described the merits of Birkenhead Park, Britain's first truly public park, in a way that prefigured what he and Vaux would do in Central Park a decade later. In 1865, having worked as a managing editor at *Putnam's Monthly Magazine*, co-founded *The Nation* and served as director of the US Sanitary Commission, he prepared *Yosemite Valley and the Mariposa Grove: A Preliminary Report*, which explained to Congress the need to preserve important American landscapes for the public. Eliot's most important work as a designer (1902) was preceded by his advocacy for public institutions to manage landscapes as shared cultural assets. His 1890 essay 'The Waverly Oaks' led to the founding of the Trustees of Public Reservations. His report two years later to the Metropolitan Park Commission educated readers about the characteristics of the Boston regional landscape to explain the need for the preservation of a particular set of sites.
14 In a recent interview with Bernd Upmeyer (2015), Jeremy Till argues that by trying to retain control and autonomy, designers trade away the ability to engage social and political processes.
15 In his essay 'Essence-less Landscape Architecture and Its Extended Family', Ian Hamilton Thompson (2013) offers an operational definition for landscape architecture that includes practices extending from high-concept design to socially driven and process-based approaches. This range should be expanded to include design work oriented not to proposing specific physical solutions but to engendering public discussion.

References

Campanella, R. (2002) *Time and Place in New Orleans: Past Geographies in the Present Day*. Gretna, LA: Pelican Publishing.
Campanella, R. (2006) *Geographies of New Orleans: Urban Fabrics before the Storm*. Baton Rouge, LA: University of Louisiana Press.
Campanella, R. (2008) *Bienville's Dilemma: A Historical Geography of New Orleans*. Baton Rouge, LA: University of Louisiana.

Campanella, R. and Campanella, C.M. (1999) *New Orleans Then and Now*. Gretna, LA: Pelican Publishing.

Colton, C.E. (2005) *An Unnatural Metropolis: Wresting New Orleans from Nature*. Baton Rouge, LA: Louisiana State University Press.

Cronon, W. (1991) *Nature's Metropolis: Chicago and the Great West*. New York: W.W. Norton.

Eliot, C. (1902) *Charles Eliot: Landscape Architect*. Amherst: University of Massachusetts Press, pp. 316–318, 384–413.

Goldberg, M.A. and Holling, C.S. (1971) 'Ecology and Planning', *Journal of the American Institute of Planners* 37(4), pp. 221–230.

Guattari, F. (2008) *The Three Ecologies*, trans. from French I. Pindar and P. Sutton. New York: Continuum.

Hamilton Thompson, I. (2013) 'Essence-less: Landscape Architecture and Its Extended Family', *Harvard Design Magazine: Landscape Architecture's Core* 36, pp. 7–14.

Hough, M. (2004) *Cities and Natural Processes: A Basis for Sustainability*. New York: Routledge.

Odum, E. (1977) 'The Emergence of Ecology as a New Integrative Discipline', *Science* 195(4284), pp. 1289–1293.

Olmsted, F. (1852) *Walks and Talks of an American Farmer in England*. New York: G.P. Putnam's Sons, pp. 74–84.

Olmsted, F. (1865) *Yosemite and the Mariposa Grove: A Preliminary Report*. Yosemite National Park, CA: Yosemite Association.

Reed, C. and Lister N. (eds) (2014) *Projective Ecologies*. Barcelona: Actar.

Sayer, A. (2011) *Why Things Matter to People: Social Science, Values and Ethical Life*. Cambridge: Cambridge University Press.

Spirn, A. (1985) *The Granite Garden: Urban Nature and Human Design*. New York: Basic Books.

State of Louisiana Office of Community Development – Disaster Recovery (2013) Greater New Orleans Urban Water Plan. Online available at: http://livingwithwater.com/blog/urban_water_plan/about/ (accessed 15 January 2016).

Thompson, J. (1957) 'The Settlement Geography of the Sacramento-San Joaquin Delta, California', PhD dissertation, Stanford University.

Till, J. (2009) *Architecture Depends*. Cambridge, MA: The MIT Press.

Upmeyer, B. (2015) 'Distributing Power: On the Complex Necessity of Participatory Urbanism', MONU: Participatory Urbanism, Autumn 2015 (#23), pp. 7–14.

Wolff, J. (2003) *Delta Primer: A Field Guide to the California Delta*. San Francisco, CA: William Stout.

Wolff, J. (2013) 'Bay Lexicon', art exhibition, Bay Observatory Gallery, The Exploratorium.

Wolff, J. (2014) 'Cultural Landscapes and Dynamic Ecologies: Lessons from New Orleans', in C. Reed and N. Lister (eds) *Projective Ecologies*. Barcelona: Actar, pp. 146–165.

Wolff, J. and Reese, C. (2009) *Ponchartrain Park + Gentilly Woods Landscape Manual*. New Orleans, LA: Longue Vue House and Gardens.

Wolff, J., Shelley, E. and Hoeferlin, D. (2011) 'Gutter to Gulf: Legible Water Infrastructure for New Orleans', online available at: www.guttertogulf.com (accessed 22 January 2016).

4

THE LAW IS AT FAULT?

Landscape rights and 'agency' in international law

Amy Strecker

Introduction

This chapter provides a critical analysis of legal approaches to landscape. It explores the potential and limits of law concerning rights *to* landscape and what they entail as well as rights *of* landscape (or 'agency') of non-human entities such as nature and the environment. Landscape is a broad, overarching concept with several nuances of meaning depending on one's discipline, professional or vocational background, and culture. In some cultures, landscape still retains an overtly aesthetic, rural connotation, while in others it is increasingly linked to the associative dimension, or non-proprietary interests in the land, or ecological values. In law, landscape was traditionally protected under cultural heritage law or natural conservation areas, both of which can be static and restrictive in their approach to human intervention. The 2000 European Landscape Convention (Council of Europe, Florence (hereafter ELC)) adopted a more democratic approach to landscape by emphasizing the central role of people and communities in the planning and management of landscapes. Discussions on landscape rights have emerged since the Convention's adoption (see for example, Egoz, Makhzoumi and Pungetti, 2011). The ELC has prompted discussions on the multiple links between landscape and democracy (see also CLAD, 2015), landscape and commons, and landscape and public space, with calls from communities and civil society groups to match. Yet the legal content of landscape rights remains unclear. This chapter discusses landscape rights and agency fifteen years after the adoption of the European Landscape Convention. It argues that the holistic concept of landscape as conceived in the ELC poses a challenge for most legal systems, because the anthropocentric structure of human rights, coupled with the elevated status of private property within national constitutions and human rights regimes, limits the possibilities for creative judicial approaches to acquiring landscape rights.

Given these limitations, and given the dynamic nature of landscape itself, it is questioned to what extent legal approaches to landscape rights can be useful.

Rights *to* landscape

When discussing landscape and human rights, the most direct route to such a discussion is through examining the relationship that various people and communities have with land outside of private property. What rights are involved in landscape aside from the most obvious right to own land? Many such rights exist, albeit not obvious at first glance. Rights to landscape may include rights of access (for example, to rights of way on public or private lands), usufruct rights (rights to fish, hunt or conduct other subsistence farming activities on public or private land); rights to enjoy sacred sites on public or private land (this is most evident in the case of indigenous communities and certain minorities); grazing rights on transhumance landscapes; rights to participate in planning decisions affecting the local landscape; and rights to a healthy environment. The latter implies the right not to have landscape damaged to the extent that it will harm human health or well-being; environmental protection per se is not at all well-established in human rights systems.

The level of recognition of the aforementioned rights to landscape varies from jurisdiction to jurisdiction. However, for the purposes of the present analysis, reference will be made mostly to the international human rights system, which supersedes national law for those states that have signed and ratified the relevant human rights treaties, and which provides an insight into the practice of various states through the case law of human rights courts. Indeed, it is through the analysis of case law that we can: (a) garner an idea of the interpretation of 'landscape' by the judiciary; and (b) what takes precedence in the hierarchy of rights. In Europe, this means examining the jurisprudence of the European Court of Human Rights, and for the purposes of comparison, the jurisprudence of the Inter-American Court. These two courts are the most prolific in terms of case law at the international level.

If the content of rights to landscape, as outlined above, includes mainly use and access, then it is those cases involving the limitation on use or access to landscapes that require attention. In addition, cases involving the challenge to development or construction in landscapes will also be addressed, as these are often based on the premise that such development will preclude the access and enjoyment of the landscape in question, and thereby infringe on certain human rights and freedoms. While the rights to landscape can be considered mostly as cultural rights, or customary rights, and in some instances also environmental rights, the human rights invoked or implied in rights to landscape are not at all obvious at first glance, and may include the right to property (in a broad sense to include custom or other rights), the right to family and private life, or rights to participate in cultural life. The lack of explicit rights to landscape makes this a necessary fact. It must also be noted that the word 'landscape' itself is not always used in the following cases.

In cases where the term 'landscape' is specifically mentioned, it is interesting to see its narrow association with preservation rather than use.

Rights to landscape in the European Court of Human Rights

The European Convention on Human Rights and Fundamental Freedoms[1] makes no reference to landscape or the environment in its text or protocols. However, rights *to* landscape (envisaged as either the right to use or access a landscape, or as 'landscape protection') can be indirectly achieved through the Convention in a number of ways: first, the protection of other rights guaranteed in the Convention (for example, the right to life) might require the safeguarding of an environment of quality. This represents an indirect form of environmental rights. Second, the right to property might entail more than mere private ownership and include other usufructuary or customary rights, such as in the case of indigenous peoples. Third, the 'general interest in a democratic society' permits restrictions on the exercise of some rights and freedoms, such as the private right to property, in favour of upholding the rights of others to access or enjoy landscapes of value. This third scenario usually involves conservationist approaches to landscape. In the realm of European case law, the question of landscape has mostly entered the Court in this way: the upholding of restrictions on individual rights in the general interest of society.

In most cases of the European Court of Human Rights dealing with landscape, landscape 'protection' (as part of environmental, cultural heritage and town planning measures) comes into conflict with other human rights and is in fact the reason for the alleged violation. Such cases usually concern complaints by persons of restrictions on the use of their property, where the Court (and formerly the Commission when it existed) have held that these restrictions have been justified for the protection of the environment, and as such necessary in the 'general interest' or for the 'protection of the rights and freedoms of others'. An early example of this was seen in the case of *Herrick v. the United Kingdom* (1985), which involved a restriction on the use of a bunker owned by the applicant on the island of Jersey.[2] The restrictive measure involved the refusal of an official permit to authorize her owning it as a summer residence. The Commission decided that the decision of the local authority was justified on the grounds of the 'general interest' to safeguard a landscape of particular interest, a green zone reputed to be one of the most outstanding features on Jersey. In upholding the restriction, the Commission stated that 'planning controls are necessary and desirable in order to preserve areas of outstanding natural beauty for the enjoyment of both the inhabitants of Jersey and visitors to the island'. Similarly, in the case of *Kozacioglu v. Turkey*, the Court upheld Turkey's conservation policies against the private claim of the owner of a property in a zone of historical and archaeological significance.[3] The Court considered 'that the protection of a country's cultural heritage is a legitimate aim capable of justifying the expropriation by the state of a building listed as "cultural property"' and further stated that 'the conservation of the cultural heritage and, where appropriate, its sustainable use, have as their aim, in addition to the maintenance of a certain quality of life, the preservation of the historical, cultural and artistic roots of a region and its inhabitants. As such, they are an essential value, the protection and promotion of which are incumbent on the public authorities.' In the recent case of *Depalle v. France*,[4] which concerned the protection of coastal zones for public use and enjoyment, the Court held that the state's interference with the applicant's right to property had pursued a legitimate aim in the general interest: to promote unrestricted public access to the shore. The Court recognized the state's wide discretion in issues concerning 'regional planning and environmental conservation policies where the community's general interest was pre-eminent'. In these cases, rights *to* landscape (conceived as right to the protection of environmental and cultural spaces) are viewed by the Court as a pre-eminent concern, and well within the scope of the state's margin of appreciation to override other individual rights and freedoms. In a way, they represent the triumph of public rights to landscape protection over private rights to property. However, in these cases the question of landscape only enters the realm of the Court in an indirect way – that is, as an external factor which impinges on the right or freedom in question. Rights to landscape were not the aim of the applicants' claims.

The second type of landscape cases to come before the European Court of Human Rights involve claims brought by ethnic minorities, such as the Roma or Irish Travellers, concerning their use and access to landscapes for dwelling purposes. Here rights *to* landscape are in a way the object of the applicants' claim, but this often comes into conflict with established landscape protection measures, thereby creating two different forms of landscape rights in direct conflict with one another: landscape preservation versus landscape use. This conflict was evident in a number of cases before the Court. In the case of *Chapman v. the United Kingdom*,[5] the applicant alleged that planning and enforcement measures against her occupation of caravans violated her right to respect for her home and private and family life (Article 8). She complained that the restrictions also entailed an interference with the peaceful enjoyment of her possessions (Article 1, Protocol 1) and that she suffered discrimination contrary to Article 14 of the Convention. The land in question was a designated 'Landscape Conservation Area' within a Green Belt zone. The applicant sought planning permission and was refused on the grounds that the occupation of her land was 'detrimental to the rural character' of the area. The Court found that the interferences did not constitute a violation of her rights under the Convention and that the decision of the local authorities was 'necessary in a democratic society'. A similar ruling was reached in the case of *Buckley v. the United Kingdom*,[6] whereby the applicant, also from the Roma community, was refused planning permission for three caravans on the grounds that the planned use of the land would detract from the 'rural and open quality' of the landscape. In the dissenting opinion of Judge Pettiti, he stated that the government's approach in this case was to give 'priority to protection of the landscape over respect for family life', thereby reversing the ranking of fundamental freedoms under Article 8. In these cases, we see a similar line of reasoning by the court as in the first type of cases above, even though the upholding of restrictions in the 'general interest' may not necessarily respect the interest of those of a different cultural or ethnic identity. The saying that landscape 'belongs to everybody' becomes problematic here, because it evidently is not everybody's landscape.

The third type of landscape cases to come before the European Court of Human Rights involve indigenous peoples' claims, and these focus on the right to use certain lands or protect them from destructive development, rather than own them in a private property sense. In relation to indigenous land rights, however, the European Court has adopted a conservative stance. First, many cases involving indigenous rights to landscape have been dismissed by the Court for procedural issues (for example, *Könkämä and 38 other Saami villages v. Sweden*; *Hingitaq 53 and others v. Denmark*).[7] Second, in cases that have been admitted, the burden to prove ancestral occupation or customary use is often onerous. For example, in an early case before the European Commission, *G. and E. v. Norway*,[8] two Sami applicants claimed that the construction of a hydroelectric dam authorized by Norway violated their property rights because the work caused the loss of part of what they considered to be their traditional lands. According to the Commission, the applicants could not provide sufficient proof of their precise connection with the flooded land, and found that consequently they had not established any property rights over the site of the dam and were not entitled to compensation. Indeed, the Commission appeared quite resistant to the idea that 'traditional use' of the land by the Sami (for hunting, fishing and reindeer grazing) could be interpreted as 'possession' within the meaning of the Convention. This refers to the wording of Article 1, Protocol 1 of the Article on the right to property: 'Every natural or legal person is entitled to the peaceful enjoyment of his possessions.' However, it was prepared to admit that the dam could affect their traditional way of life in the area in a way that triggered the application of Article 8 (right to family and private life). Interestingly, the Commission later reversed its view on possession by stating in relation to another case that Sami hunting and fishing rights could be 'regarded as possessions within the meaning of Article 1, Protocol 1 of the Convention'.[9] In a more recent case involving an indigenous right to landscape made under Article 1, Protocol 1, *Handolsdalen Sami Village v. Sweden*,[10] the Court was asked to rule on the admissibility of winter grazing. The case was brought because the Sami applicants had lost their case in Swedish courts to have their traditional winter grazing rights recognized on land belonging to private parties. The applicants argued that this violated their rights under Article 1, Protocol 1. However, the Court found that the right to access private property for the purposes of reindeer grazing had no basis in law and was inadmissible. In her dissenting opinion, Judge Ziemele referred to the developments in international law concerning indigenous rights and challenged the Court's decision. Other authors have criticized the European Court of Human Rights for its conservative approach to indigenous land rights cases.[11] This stands in contrast to the approach of the Inter-American Court and Commission, which have elaborated a significant case law recognizing indigenous customary rights to lands despite lack of title (to be discussed below).

What the aforementioned cases illustrate is that the European Court of Human Rights is rather conservative in the realm of rights to landscape, except when those rights entail landscape protection or preservation in the 'general interest of society'. This implies two things: first, landscape protection, as decided by local authorities, is considered important enough and within the state's margin of appreciation to override other rights and freedoms. Yet this is still a negative form of protection, meaning that landscape preservation was not the aim of the applicants, but the problem. Second, non-traditional forms of property rights, such as rights of use or access, are not seriously considered by the Court, neither in the case of minorities nor indigenous peoples. This represents a scenic, preservationist approach to landscape. Of course, part of the reason is due to the lack of specific mention of cultural rights in the Convention, in addition to the fact that the right to property does not have collective but rather individual connotations and is restricted to the enjoyment of one's possessions, despite the Commission's pronouncement that it can also include hunting and fishing rights.

Rights to landscape in the Inter-American Court of Human Rights

The case law of the Inter-American Court of Human Rights is significantly more progressive than the European Court of Human Rights in matters of rights to landscape. This is for a number of reasons. First, most of the cases dealing with the rights to access or use lands, or to have them protected from destructive development, are based on claims by indigenous peoples, who have a special status under some national constitutions in the Americas, and which the Court officially recognizes. The Americas are home to much larger populations of indigenous peoples than Europe. At the same time, however, due to the resource-rich nature of their traditional lands, especially in the Amazon region, conflicts often arise between logging, mining and hydroelectric projects approved by the state, and local communities, which is the reason for so many cases in this matter. As noted by the UN Committee on Economic, Social and Cultural Rights, the traditional lands of many indigenous peoples have been reduced or occupied, without their consent, by timber, mining and oil companies, at the expense of their culture and the balance of the ecosystem.[12] This represents a continuous form of repression and dispossession under the guise of 'development', and is reminiscent of the *terra nullius* doctrine which was used by states to legitimatize colonization, declared by the ICJ as far back as 1975 as 'erroneously and invalidly applied'.[13]

Many cases concerning rights to landscape in the Inter-American system concern the challenge to the granting of logging or mining concessions on communal or ancestral lands, where a community lacks proper title but has traditionally occupied the area. In these cases the interventions by large-scale agro-forestry projects would have a severe impact on the communities' way of life. For example, in the case of *Awas Tingni v. Nicaragua*,[14] a Korean corporation was granted concessions by Nicaragua in 1996 to commence logging in the communal lands of the Awas Tingni community. The community tried to prevent the government from proceeding, first through negotiations and then by resorting to the judiciary. The case was eventually brought before the Inter-American Commission on Human Rights and then before the Inter-American Court of Human Rights. In a landmark decision, the Inter-American Court held that the international human right to property, particularly as affirmed in the American Convention on Human Rights, includes the right of indigenous peoples to the protection of their customary land and resource tenure. The Court held that Nicaragua violated the property rights of the Awas Tingni community by granting a foreign company logging concessions within the community's traditional lands and by failing to otherwise provide adequate recognition and protection of the community's customary tenure. As pointed out by Anaya, this was the first legally binding decision by an international tribunal to uphold the collective land rights of indigenous peoples in face of the state's failure to do so.[15] Likewise, in *Maya Indigenous Community of Toledo v. Belize*,[16] the Inter-American Commission found that Belize had violated the Mayan communities' right to use and enjoy their property by granting concessions to third parties to exploit natural resources within Mayan lands without informed consent. The Commission noted that indigenous peoples' right to property is based in international law and does not depend on domestic recognition of property interests. The Commission broadly defined indigenous property rights as not limited 'exclusively by entitlements within a state's formal legal regime, but also include that the indigenous communal property that arises from and is grounded in indigenous custom and tradition' and further stated that 'the distinct nature of the right to property as it applies to indigenous people, whereby the land traditionally used and occupied by these communities plays a central role in their physical, cultural and spiritual vitality'.[17]

The second form of indigenous rights to landscape cases to come before the Inter-American system concern the protection of other rights guaranteed in the Convention (for example, the right to life) that indirectly require the safeguarding of the landscape. Indeed, as early as 1985, in the case of *Yanomami v. Brazil*, the Inter-American Commission established a link between environmental quality and the right to life in response to a petition brought on behalf of the Yanomami of Brazil.[18] The petition alleged that the government violated the rights of the Yanomami by constructing a highway through their territory and authorizing the exploitation of the territory's resources. These actions led to the influx of non-indigenous people who brought diseases which remained untreated due to lack of medical care. The Commission found that the government had violated a host of rights including the right to life, liberty and personal security, as well as the right of residence and movement, and the right to health and well-being. In the more recent case of *Moiwana Community v. Suriname*,[19] the Inter-American Court dismissed Suriname's argument that the right to property of an indigenous community was time-barred. It concluded that in cases where the state is directly responsible for the original displacement of the population and does not provide for their safe return, the right to claim collective property is not lost. The village of

Moiwana remained abandoned after a 1986 attack by government forces, who killed thirty-nine unarmed members of the community, including women and children. The community's property was destroyed and survivors were forced to flee. Throughout the case, particular emphasis was placed on the special relationship between the N'djuka and their land, and the consequent violation of rights arising out of their displacement. The Court recalled that 'in case of indigenous communities that have occupied their ancestral lands with their customary practices – but lack of a formal property title – the possession of the land should be sufficient to obtain the official recognition of the property and to obtain subsequent registration'.

The third type of landscape case to present itself before the Inter-American Court concerns the aforementioned conflict involving landscape preservation versus landscape use. In a recent decision in 2010, the Inter-American Court ruled against Paraguay for failing to guarantee the communal right to property of the Xákmok Kásek Indigenous community.[20] The community began proceedings as far back as 1990 with the aim of recovering their ancestral lands. In 2008, the government declared 12,450 hectares of the Salazar estate a Protected Wildlife Area under private ownership (Decree 11804/08), without consulting the community members or considering their territorial claim, despite the fact that 4,175 hectares of the estate overlapped with lands being claimed since 1990. In its decision, the Inter-American Court found that throughout these years of struggle, the Xákmok Kásek community suffered significant restrictions on the use of their land by the private owners of the estate. These restrictions intensified in recent years, as the community found its traditional subsistence activities and movement around the lands increasingly limited. Hunting was banned and the private owners hired guards to monitor activities, so that gathering food and fishing became impossible. In its decision, the Inter-American Court found that Paraguay had violated the communal right to property, the right to life, the right to personal integrity and the rights of the child. Notably, the Court reiterated its consideration in previous decisions that the 'close relationship of indigenous peoples to their traditional lands and the natural resources relevant to their culture that are found there, as well as the intangible elements resulting from them, must be safeguarded under Article 21 of the American Convention'.[21] It also noted that the concept of property in the indigenous context can have a collective meaning, in the sense that possession is not focused on individuals but on the group and its community, and that while this concept of property does not necessarily correspond to the classic concept of property, it nevertheless deserves equal protection under the Convention. Indeed, the Court further expanded on its interpretation of property by stating that 'failing to recognize the specific versions of the right to use and enjoyment of property would be equivalent to maintaining that there is only one way of using and enjoying property and this, in turn, would make the protection granted by Article 21 meaningless for millions of individuals'. The significance of this ruling for the recognition of the landscape rights of indigenous peoples cannot be underestimated. The Court has firmly acknowledged that the concept of property entails several aspects, and not solely the fact of abstract title. This essentially represents a right to landscape, or a right to property as originally conceived, before the term evolved to connote mere title or *ownership* of the property rather than the *identification* of the individual with the property through custom.[22]

From the cumulative body of case law of the Inter-American Court dealing with indigenous rights to landscape, it is evident that: (a) the Inter-American Court adopts a much broader interpretation of property than its European counterpart. This broad interpretation includes a collective, customary and intangible dimension that is linked to a way of life; and (b) these collective rights are considered more important than landscape preservation. There is also a strong cultural element in these cases which is absent in the case law of the European system and this is partly due to the existence of a right to culture in the American Convention (Article 14, Additional Protocol), as well as a right to healthy environment (Article 11, Additional Protocol).

Rights *of* landscape in Human Rights Courts

The first point to note when discussing rights of landscape in human rights terms is that they are immediately limited by the anthropocentric nature of human rights, despite the dependence of humankind on a functioning healthy environment. As seen in some of the cases above, landscape protection has been advocated when it affected other human rights and freedoms, but was not the subject of the claims in and of itself. It was therefore a 'negative' form of protection: safeguarding the landscape was not the aim of the applicants. Cases in international human rights courts arguing for landscape protection are few and far between, but there are some examples. However, the lack of any explicit rights or 'agency' of non-human entities such as nature or the environment makes these cases extremely unlikely to succeed. Even in the Inter-American system, where the right to a healthy environment is explicitly recognized in the Additional Protocol to the Convention, the emphasis is indeed on the *human* right to a healthy environment, and not the environment's right to be healthy.

Rights of landscape in the European Court of Human Rights

As previously stated, there is no mention of a right to a healthy environment in the European Convention on Human Rights, nor is there any explicit reference to a right to culture or heritage. However, this has not precluded the attempt by some applicants to bring cases dealing with landscape protection before the court. A case in point is that of *Kyrtatos v. Greece*,[23] which concerned a tourist development in a wetland of Ayios. In 1993 the applicants and the Greek Society for the Protection of the Environment and Cultural Heritage applied to the Council of State for judicial review of illegal building permits. Their main argument was that they were illegal because there was a wetland in the area concerned and, under a constitutional provision protecting the environment, no buildings could be erected in an important natural habitat for various protected species. The applicants complained that the destruction of the wetland adversely affected their lives, even though it posed no danger to health. The Court did not accept that the interference with the conditions of natural life in the area constituted an attack on the applicants' private or family life (Article 8). So while the building permits and consequent destruction of the wetland resulted in a violation of the law, the applicants could not show how the alleged damage to the environment had directly affected their rights under the Convention. The Court reaffirmed in this regard that neither Article 8 nor any of the other Articles of the Convention are specifically designed to provide general protection of the environment as such.

Representative actions seeking to challenge the law in the abstract are not accepted by the ECtHR.[24] As the Court pointed out in another case, 'the applicant cannot complain as a representative for people in general, because the Convention does not permit such an *actio popularis*. The Commission is only required to examine the applicant's complaints that he himself was the victim of a violation.'[25] Thus the ECtHR's approach to standing is quite narrow. Article 34 of the Convention, which provides for applications from non-governmental organizations or group of individuals, still requires that the applicants be victims of an alleged violation and prove that they are directly affected by the matter complained of. In the case of associations and NGOs, they must prove that they are themselves in some way affected. Even clear violations of the rule of law cannot be remedied if the applicant is not sufficiently affected or if no direct link between the alleged victim and the violation can be proven. As pointed out by Schall, the substantive law available in the ECHR is anything but ideal for public interest proceedings.[26] The strict approach of the Court in matters of the environment would therefore preclude a challenge on landscape grounds because: (a) environmental integrity is not seen as a value per se for the community affected or society as a whole but rather as a criterion to measure the negative impact on a person's life, property, private and family life; and (b), the individualistic approach followed by the Court excludes the admissibility of public interest proceedings on environmental grounds, unless the applicants can show a direct impact of the activities complained of in relation to their individual rights. Thus resources that are more widely shared (such as landscapes) cannot be protected by the European Convention, even if the applicants, as in *Kyrtatos*, have a special interest in that particular part of the environment.

Rights of landscape in the Inter-American Court

While all of the above cites cases from the Inter-American system concerned the rights of indigenous peoples, in *Metropolitan Nature Reserve v. Panama*,[27] the applicant filed a petition to the Inter-American Commission on behalf of the citizens of Panama. He alleged that the construction of a pubic highway through the Metropolitan Nature Reserve violated the Panamanian citizens' rights to property, in so far as the Nature Reserve is designated as a protected area of environmental, scientific and cultural value for all Panamanian people by law. The Inter-American

Commission concluded that the petition was inadmissible as it was overly broad and the applicant had failed to identify victims. This approach of the Commission can be contrasted with the cases above concerning indigenous rights to communal lands. In the former cases, the right to property was construed by the Court and Commission to include the customary right of indigenous communities to access and use the lands they had occupied but without title. Apart from the central role these lands play in the 'cultural and spiritual vitality' of indigenous peoples, there is still very much a use and dependency value. In the latter case, however, the issue concerned a nature reserve of environmental and cultural value for the 'citizens' of Panama, and not for one particular group. The difference between the former cases and the latter is emblematic of the problem that exists when articulating landscape protection in rights language. In the Inter-American system, a collective approach – and a broader interpretation of property – is adopted, but only in relation to indigenous peoples. When it involves a case dealing with a nature reserve or public space, the approach is very different. This is the same in the European context, where the narrow conceptualization of rights and the required level of standing preclude the admission of public interest proceedings, even if the area is protected by law and is of significance for the citizens of the state.

Conclusion

To conclude, this chapter has provided a critical analysis of international case law dealing with rights *to* landscape and what they entail, as well as rights *of* landscape (or agency) of non-human entities such as nature and the environment. Given the anthropocentric nature of the human rights framework, most of the case law has dealt with the former. It can be seen from the foregoing analysis that the approaches of the two main human rights courts diverge in this regard. The European Court of Human Rights adopts a conservative approach in the realm of rights to landscape, except when those rights entail landscape protection or preservation in the 'general interest of society'. Yet this is still a negative form of protection, meaning that landscape preservation or access was not the aim of the applicants, but the problem, and it consequently entailed a limitation on other rights and freedoms. Second, non-traditional forms of property rights, such as rights of use or access, are not seriously considered by the European Court. This contrasts greatly with the Inter-American jurisprudence on indigenous rights to landscape, which has developed in recent years to incorporate a more collective, customary and intangible dimension to property that is linked to a way of life. There is also a strong cultural element in these cases which is absent in the case law of the European system and this is partly due to the existence of a right to culture in the American Convention (Article 14, Additional Protocol), as well as a right to healthy environment (Article 11, Additional Protocol). However, this broad approach is only associated with indigenous communities and not the body of citizens as a whole. In relation to rights *of* landscape, the limited available case law reveals that unless the landscape degradation under question has a direct impact on an individual's other human rights and freedoms (such as right to life) then cases are unlikely to succeed. The lack of any explicit rights or 'agency' of non-human entities such as nature or the environment makes this a reality. Even in the Inter-American system, where the right to a healthy environment is explicitly recognized in the Additional Protocol to the American Convention, the emphasis is indeed on the *human* right to a healthy environment, and not the environment's right to be healthy. The view that mankind in part of, as opposed to separate from, a global ecosystem may reconcile the aims of human rights and environmental protection. Indeed, at the national level, some courts have begun to broaden standing requirement to allow legal redress for violations of environmental rights without necessarily requiring individual injury to health or property, because one major motive for guaranteeing environmental rights is to prevent injury from occurring.[28] A progressive provision has recently been included in the Ecuadorian Constitution, which declares nature a legal person in Ecuador.[29] The Constitution states that 'the government has the duty to protected nature on behalf of all the inhabitants of Ecuador present and future' and that 'nature has the right to exist, persist, maintain and regenerate its vital cycles, structure, functions and its processes in evolution'. Furthermore, 'should governments fail in this duty at any level, any citizen of Ecuador can still on behalf of nature force the government to take appropriate action'. This is essentially a provision granting 'agency' to nature. Whether and to what extent this provision will be used in practice remains to be seen. Lastly, in the absence of an international environmental court, human rights courts offer one of the only possibilities for citizens to challenge governmental decisions and attempt to control the abuse of power by the state. The creative approach by the Inter-American Court offers an example of creative judicial activism, which could be adapted and applied to other contexts if similar conditions prevailed. A broader interpretation of the content of property rights to include custom, use and access, is welcomed and need not be restricted to indigenous peoples, but be adaptable to other communities and groups in relation to landscape rights.

Notes

1. Council of Europe, *European Convention for the Protection of Human Rights and Fundamental Freedoms, as amended by Protocols Nos. 11 and 14*, 4 November 1950, ETS 5.
2. Application No. 11185/84, DR 42, 275.
3. *Kozacioglu v. Turkey*, Grand Chamber decision 19/02/09. Application No. 2234/03.
4. *Depalle v. France*, Grand Chamber judgment 29/3/2010. Application No. 34044/02.
5. *Chapman v. the United Kingdom*. Grand Chamber judgment 18/01/2001. Application No. 27238/95.
6. *Buckley v. United Kingdom*, Judgment 29/09/1996, Application No. 20348/92.
7. *Konkäma and 38 Sami villages v. Sweden*, Application No. 27033/95, Decision 25/11/1996. *Hingitaq 53 and Others v. Denmark*, Application No. 18584/04, Decision 12/01/2006.
8. *G. and E. v. Norway*, Application No. 9278/81 & 9415/81 (joined). Decision of 03/10/1983.
9. *Könkäma*, supra note 8, at 85.
10. *Handölsdalen Sami Village and Others v. Sweden*, Application No. 39013/14. Judgment of 30 March 2010.
11. See for example, Otis, G. and Laurent, A. (2013) 'Indigenous Land Claims in Europe: The European Court of Human Rights and the Decolonization of Property', *Arctic Review on Law and Politics* 4 (2/2013), pp. 156–180.

12 Concluding observations of the committee on economic, social and cultural rights Columbia, UN Doc.E/C.12/1/Add.74 of 30 November 2001, paras. 11 and 12.
13 *Western Sahara* Case, ICJ Reports (1975), 12.
14 *The Mayagna (Sumo) Awas Tingni Community v. Nicaragua*, Judgment of 31/08/2001, Inter-Am. Ct. H.R., (Ser. C) No. 79 (2001).
15 Anaya, J. and Grossman, C. (2002) 'The Case of Awas Tingni v. Nicaragua: A New Step in International Law of Indigenous Peoples', *AJICL* 19(1), pp. 1–15.
16 *Maya Indigenous Communities of Toledo District v. Belize*, Case 12.053, IACtHR Report 40/04 (2004) at 153, 194.
17 Ibid., at p. 155.
18 *Yanomami Case*, Case 7615, Inter-Am. C.H.R. Res. No. 12/85, OEA/Ser.L/V/II.66, doc. 10, rev. 1, 24 (1985).
19 *Moiwana Community v Suriname* (IACrtHR) Judgment of 15 June 2005, Series C No. 124.
20 *Case of Xákmok Kásek Indigenous Community v. Paraguay*, IACtHR Series C No. 214.
21 See para. 85, where specific reference was made to: the *Case of the Yakye Axa Indigenous Community v. Paraguay*, Judgment of 17 June 2005, Series C No. 125, para. 137; the *Case of the Sawhoyamaxa Indigenous Community v. Paraguay*, Judgment of 28 November 2007, Series C No. 172, para. 118; and the *Case of the Saramaka People v. Suriname*, Judgment of 29 March 2006, Series C No. 146, para. 88.
22 For a discussion of the origins of property, see Graham, N. (2010) *Lawscape: Property, Environment, Law*. London: Routledge, especially p. 26.
23 *Kyrtatos v. Greece*. Application No. 41666/98, Judgment of 22/05/2003.
24 Persons, non-governmental organizations or groups are allowed to bring a case if they can prove they are victims of a violation of one of the rights enshrined in the Convention (Article 34).
25 *X Association v. Sweden* (1982) 28 DR 204, 206.
26 Schall, C. (2008) 'Public Interest Litigation Concerning Environmental Matters before Human Rights Courts: A Promising Future Concept?', *Journal of Environmental Law* 20(3), p. 428.
27 *Metropolitan Nature Reserve v. Panama*, Case 11.533, Report no 88/03, IACtHR, OEA/SerL/V/II.118 Doc 70 Rev 2 at 524 (2003).
28 See for example, the case *Montana Environmental Information Centre v. Department of Environmental Quality*, 296 Mont. 207, 988, P.2d 1236 (1999).
29 Articles 86 through 91.

5

HOW TO LIVE IN A JUNGLE

The (bio)politics of the park as urban model

Maria Shéhérazade Giudici

I

Making what is cold warm, what is humid dry, what is dark light: architecture has always strived to control climate, albeit in a rather clumsy and limited way. However, *designing* climate – which would have once been a feat of magic – has now become technically possible and is an increasingly fundamental aspect of the contemporary work of architects. The applications of these techniques to interior spaces are well known, so much so that architects often take for granted the idea that the post-modern city has become nothing but an immense air-conditioned interior.[1] On the other hand, only a few projects have focused on the possibilities of climate control when it comes to open-air spaces, at least until recently. This is what Swiss architect Philippe Rahm and French *paysagiste* Catherine Mosbach[2] are currently trying to do in Taichung, a city of 2.7 million people on the western coast of Taiwan.[3] The ground for this experiment is the Jade Eco Park, which the architects conceived as a series of haptic experiences rather than a traditional composition of vistas. Humidity, temperature, wind condition here become as important – indeed, more important – than colour and shape. The designers argued for the creation of microclimates as a smart solution to Taichung's warm and humid subtropical climate which would normally impede the use of the park as a public space as happens in more temperate climates.[4] The ideas underlying the Jade Eco Park are not merely functional, however, and the project can also be read as an ideological statement against the long-standing architectural tradition of addressing sight before other senses. In this rediscovery of the non-visual, of the haptic, of the physical qualities of architecture there is a definite attempt to find new forms of agency – that is to say, of ways for the individual to act within and against his/her context.

This project, however, does not only deal with an idea of individual agency at the scale of our body, but also with the opportunity for architecture to retain a possibility of agency on the scale of the city. Few recent urban plans have presented such a clear ambition to say something about what a city should be. Ever since the collapse of the welfare state and the triumph of neoliberal urban politics, city administrations and architects alike have by and large bailed on the possibility to put forward comprehensive projects for the city. Even when new cities are still being planned from scratch in developing countries, very often they result in plans aimed at strategically managing risk and growth rather than putting forward a precise idea of the urban and its subject. As architecture seems to have abandoned any hope of direct agency on the large scale, very little experimentation has happened recently in terms of urban design; however, there have been in the last few decades a number of proposals for *parks* that go beyond the cosmetic and propose rather radical urban scenarios. The Paris La Villette Park competition of 1982 marked the first official acknowledgement of the death of the industrial city in the West, for instance[5] – just as much as the 1999 Toronto Downsview Park proposed the predominance of process over form, and the 2011 Jade Eco Park perhaps moved urban design outside of the visual realm. Already at the stage of the competition brief,[6] these projects were clearly instrumental for the respective city administrations in the same way in which in the heyday of the welfare state a social housing project would have been. The park, which had long been the space of Arcadia, of leisure, of the 'other', has become today the primary testing ground for urban models.

Arcadia is the promise of a peaceful coexistence of man and nature – first and foremost, an ideological lie: any peace can only be a truce of which man is the temporary winner. Indeed, Arcadia is the most unnatural of all possible conditions, a mere narrative device invented to serve in turn as critique or apology of reality. If architecture was born as a primitive way to protect ourselves from nature – and, later, tame nature – Arcadia is the very opposite of architecture, a world in which architecture is not needed. And yet, *et in Arcadia ego*: there is, arguably, no nature untouched by architecture, not even Poussin's idyllic countryside, dominated by a classical tomb. On the other hand, in the last two centuries one could very well reverse the saying, and argue that Arcadia itself has managed to rise to prominence as one of the most pervasive figures of both architectural debate and city-making at large. As much as the temperature-controlled *city-as-interior* trope is a fundamentally correct portrait of the contemporary metropolis, one could challenge it by saying that the urban condition has, at the same time, become a continuous exterior, designed as a park, and meant to let us behave as we would in a field, a forest, a jungle. In fact, in a 1955 interview, none other than Mies van der Rohe categorically stated that 'There are no cities, in fact, anymore. It goes on like a forest. . . . We should think about the ways we have to live in a jungle, and maybe we do well with that.'[7]

In this context, the ability of the individual – be it the architect or the citizen – to act,[8] to react, or to simply be aware, is dissolved by the apparent lack of structure. The concept of agency is dialectically linked to the presence of a structure – social, formal or spatial;[9] but the hallmark of the city-as-a-forest is its apparent lack of structure, its pretended 'naturalness'. The following pages will argue that paradoxically, in this condition, the project of the park has become one of the last occasions of agency

How to live in a jungle 67

for architects at the scale of the city. By re-discussing the Jade Eco Park and comparing it with OMA's Downsview Park proposal, I will also try to see if, and how, a park project today can question and challenge the city-as-a-forest from within.

II

The Jade Eco Park is the product of the collaboration between *paysagiste* Catherine Mosbach, an expert of landscape design, and Philippe Rahm, an architect whose work revolves around the idea of making space through the creation of different body conditions. In his projects of various scale – from installations to urban proposals – Rahm has consistently explored the possibility of giving a rhythm to space through the manipulation of atmospheric parameters rather than through walls and traditional architectural elements. The Jade Eco Park can be considered the *summa* of these experiences as it is one of the most recent, and perhaps the largest project, realized by Rahm. On a 70-ha area that formerly hosted an airstrip, the designers imagined a sequence of spaces that vary from dry to humid, from warmer to cooler, from more exposed to more protected from urban pollution. These atmospheric conditions are achieved with a mixture of natural and technological elements; on the one hand, the project presupposes a careful reading of the conditions of the existing site such as wind, topography and tree species, and on the other, it enhances and corrects these conditions with devices that further cool and dehumidify the air. The park thus conceived manages to be at the same time a cutting edge, highly technological architectural tour-de-force, but also a landscape modelled as a 'natural' space.

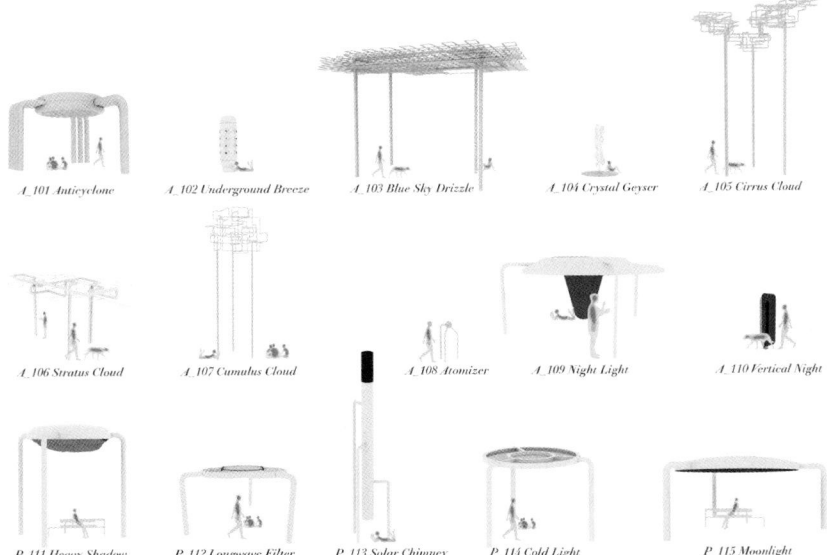

FIGURE 5.1 Catalogue of climatic devices, Jade Eco Park, 2012–2016, Taichung, Taiwan, Philippe Rahm architectes, Mosbach paysagistes, Ricky Liu & Associates.

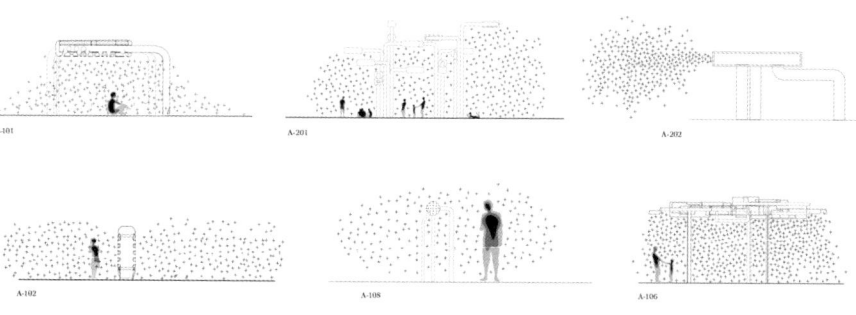

FIGURE 5.2 Cooling climatic devices, Jade Eco Park, 2012–2016, Taichung, Taiwan, Philippe Rahm architectes, Mosbach paysagistes, Ricky Liu & Associates.

The theme of landscape and of the interaction between natural and man-made environment had been at the centre of Rahm's interest long before this project, ever since his collaboration with Jean-Gilles Décosterd in the late 1990s. In 2002, Décosterd and Rahm presented at the Venice Biennale an installation titled *Hormonorium*, in which the condition of an alpine sunny day was recreated not visually, but through the deployment of a series of physiological strategies that would make the visitor *feel* as if s/he were standing on a mountain at 3000 m altitude.[10] The actual architectural composition of the room was extremely simple and not visually reminiscent of a natural site – in fact, it was just a white room with a glowing platform at its centre. However, the physical conditions recreated in the room would simulate very closely a high-altitude environment therefore triggering a body response similar to that one would experience at high altitude. The level of oxygen was lowered from 21 per cent to 14.5 per cent, and the light came from a white acrylic floor that reproduced the condition of reflected light one would experience if walking on snow. The *Hormonorium* underscored the importance of our physical reaction to space, a reaction that goes even beyond the senses to influence our endocrine and neurovegetative systems. In fact, the lowered oxygen content and effect of reflected radiation can cause dizziness, disorientation but also a sense of euphoria and changed mood in the visitor.

At the 2008 Biennale Rahm presented another installation that went even further in the exploration of our bodily understanding of space; *Digestible Gulf Stream* was composed of two platforms, a warm one positioned at low height, and a cool one at a higher level.[11] The relative position of the two devices generated an air current from warm to cool through natural convection, therefore creating an interior landscape of sorts whose functional interpretation was left open to the user. This notion of interior landscape present both in the *Hormonorium* (which recreated a real landscape) and in the *Gulf Stream* (which organized an interior with the logic of a natural exterior) indeed returns in many of Rahm's projects which are examples of 'meteorological' conditions being recreated through minimal, sometimes molecular moves within technologically sophisticated architectures.

In Taichung, Rahm and Mosbach have the chance to work on an actual large-scale environment which has to include a design for plants and trees. In Rahm's other projects architecture is always present as a frame of the intervention itself, retaining a clearly man-made character, but in the Jade Eco Park architecture transforms into the art of orchestrating environments in its purest, clearest version. The project is literally conceived as a series of environmental pockets whose differences could be entirely described as body conditions. Three landscapes – a landscape of heat, one of humidity and one of pollution – overlap creating a variety of spaces. Dehumidification, dryness and cleaner air are achieved both through the planting of carefully selected species, and through custom-designed machines.

If it is true that projects such as the *Digestible Gulf Stream* had already targeted inhabited space as the habitat of the human 'animal' body, in the case of the park the slippage presented by the artificiality of the previous installation is absent and therefore the ideological implications of Rahm's architecture become clearer.

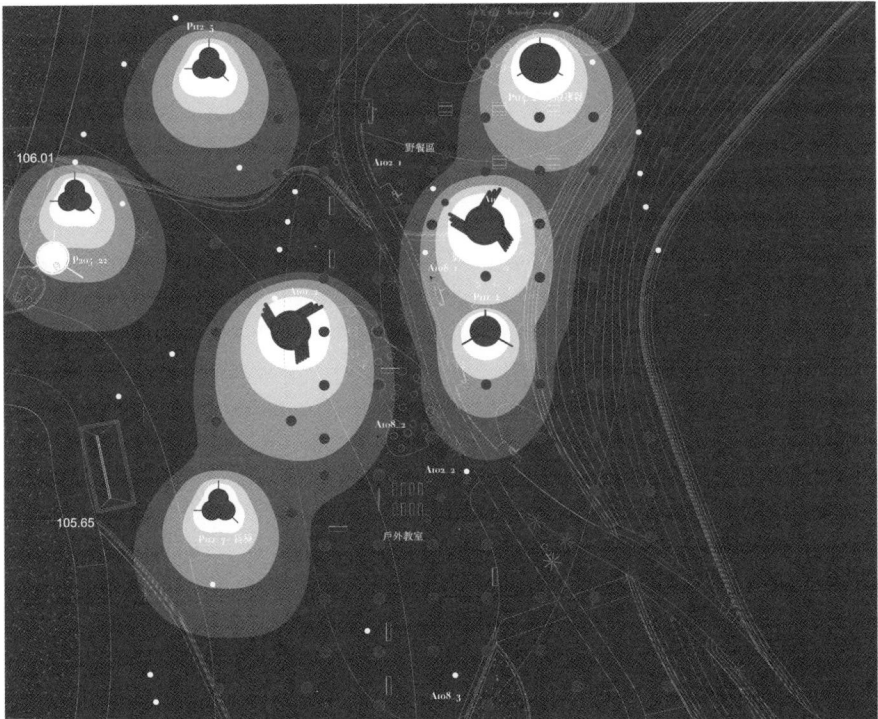

FIGURE 5.3 Detail of the plan with effect of climatic devices, Jade Eco Park, 2012–2016, Taichung, Taiwan, Philippe Rahm architectes, Mosbach paysagistes, Ricky Liu & Associates.

In fact, what is most fascinating in the atmospheric architecture of Philippe Rahm is the extreme reduction of the performance of space to a sensorial dimension. This sensorial dimension acts on the least culturally formed of our senses – touch – as well as on our internal chemistry and physiology. By marginalizing the importance of sight, the user is encouraged to shed a whole baggage of pre-received cultural ideas as s/he is called first of all to respond with his/her body rather than his/her intellectual and historically formed understanding. There is something profoundly liberating in this possibility and this seems indeed to be Rahm's intention, as he has stated that he hopes his design method can ultimately encourage alternative kinds of use of space and social imagination.[12]

Of course, any interaction between man and environment is always somehow constructed and culturally mediated even when it does not involve the sense of sight. What we perceive as too cold or too warm can change drastically from a culture to another, for instance. However, by and large the last century, the century of the controlled environment, has flattened these differences and the haptic is still by far the sense that has remained the least subject to cultural manipulation. To revert to the haptic means to go back to basics, to put the body at the centre of the project – to reclaim the body's very agency.

The idea that our ultimate liberation from the constraints of history and economy could come through our body is not a new one,[13] but Rahm's take on the subject is very extreme as it is based on an understanding of what our body can perceive that is well beyond traditional architecture. This strategic exodus of architecture into the molecular and the meteorological[14] raises some interesting questions. Although we often perceive the role of the body as diminished in our increasingly 'virtual' existences, one could also argue the opposite – that is to say, that we are, more than ever before, reduced to mere *countable* bodies. Indeed, this is one of the key principles of the biopolitical era: gone is the traditional understanding of man as a soul, enter the statistic concept of body as receptacle of productive potential (male, female, rich, poor, Caucasian or Asian and so on). It is perhaps a step forward in equality, but, as Foucault would say, to a condition of more widespread empowerment corresponds also a more refined form of enslavement.

Biopolitics is about micromanaging life, starting indeed from the body and therefore the realm of reproduction. From this point of view, Rahm's work on the design of internal choreographies of hormones and physiological responses takes to its extreme consequences the conditioning role that architecture has assumed in the age of biopolitics. However, the *bios* is not mere bodily life, and in fact micromanaging our lives means increasingly to micromanage our cultural expectations and desires. It is in this discrepancy between body and culture, *zoe* and *bios*, that Rahm's atmospheric architecture finds a new possibility for the agency of the individual user. As the cultural dimension is stripped bare, architecture becomes pure conditioning of bodies – it is a condition so radical, that it challenges all the social and moral categories that biopolitics is so intent on constructing. As the user becomes pure animal body, opportunity of misinterpretation, misuse and escape open up.

How to live in a jungle 71

FIGURE 5.4 Functioning stratus cloud device on site, Jade Eco Park, 2012–2016, Taichung, Taiwan, Philippe Rahm architectes, Mosbach paysagistes, Ricky Liu & Associates.

FIGURE 5.5 Stratus cloud cooling device on site, Jade Eco Park, 2012–2016, Taichung, Taiwan, Philippe Rahm architectes, Mosbach paysagistes, Ricky Liu & Associates.

Moreover, if it is true that the Jade Eco Park deals with the body, the most 'natural' part of us, it also subtly undermines the very meaning of nature as the whole system is in fact very carefully scripted and achieved through the use of technical devices. Machines called 'anticyclones' blow cool air to create pleasant pockets of cool temperature; cooling is also achieved by body contact of visitors with special black surfaces that are chilled with water and by vaporizing water in devices called 'stratus cloud' and 'blue sky drizzle' as well as reflective apparatuses that deflect the presence of direct sunlight.[15] Drying devices and antipollution stations complete the furnishing of the park to the point that it is almost impossible here to distinguish between nature and machine. The explicit artificiality of the Jade Eco Park is what makes it ultimately so interesting. On the one hand, it does reduce the visitor to a body, that is, to an animal – but on the other hand, it clearly showcases the fact that the environment in which this animal lives is not at all 'natural'.

The environmental work of Rahm poses several questions not only to the discipline of architecture, but to our political imagination at large.[16] Up to what point does the dialectic of artificial and natural still make sense today? How does the use of nature and natural environments impact city-making in an advanced neoliberal scenario? How does contemporary capitalism deal with the paradoxes of a subject that is at the same time more 'animal' than ever, while becoming increasingly virtual?

Ultimately, is the body our prison, or our escape route?

III

Rahm and Mosbach's proposal does not put forward an idea of park that is antagonistic to the city; indeed, one could imagine applying the same working method to the city at large and designing an urban environment as a climatic or atmospheric project at large.[17]

If the economic and social potential of the park had already been clear to city-makers for a couple of centuries, the idea of establishing the park as a formal *model* for the urban is a rather recent one, since the first wave of public parks created around the mid-1850s proposed, on the contrary, parks that were alternative if not downright antagonistic in relationship with the city. One of the clearest examples of this tendency is Alphand's system of parks for Paris, designed under Haussmann's tenure.[18] Alphand systematically replaced all but two of the straight *allées* that had marked the Bois de Boulogne with curvilinear paths inspired by picturesque models and proceeded to design in the same language several other smaller gardens as well as the large Parc de Vincennes. Alphand's parks were meant to complement the straight avenues of Haussmannian Paris; by turns heterotopian laboratory and compensatory fantasy, natural urban space was in fact a fundamental part of the very idea of modern city and its role was ideological far before being pragmatic. As Alphand rightly interpreted it, throughout antiquity up until the industrial metropolis, the park was in fact instrumentalized as the space of *otherness*.

This otherness had to do with the distinction between tame park and wild nature first, and later with the contrast between clean 'nature' and dirty industrial city. However, almost two centuries afterwards, we are now in an opposite situation and the park is no longer an enclosed idiosyncratic condition, but is rather conceived as the ultimate (though unattained and perhaps unattainable) possible configuration for the city. The recent architectural competition that more than any other has embodied this attitude was launched in 1999 for the Downsview

Park in Toronto. Downsview, much as the Jade Eco Park site, hosted a former airstrip as well as a military base that at the end of the 1990s was in need of ideas for renewal. At 230 ha, the park is roughly three times as big as the Taichung area and can therefore be considered a proper piece of city containing five actual districts and many buildings, both commercial and residential. Five entries were shortlisted, with OMA's proposal eventually selected as winner, but more than fifteen years later the scheme still hasn't been implemented.[19]

The competition brief put forward a specific idea of process-based design and encouraged the collaboration between specialists of different fields, pushing the concept of landscape urbanism as a large multidisciplinary umbrella that would cover both park design and urban design. In fact, while the area is called a park, it is questionable whether we are looking at a park project or rather at a low density urban development. Both brief and proposals stressed the importance of an economic scenario for the park, planning it as a value-producing machine in and of itself. Here the park is seen as *active* part of a productive tissue and therefore, again, in fundamental continuity with the city.

The OMA statement on the park reads:[20]

> Trees rather than buildings will serve as the catalyst of urbanization. Vegetal clusters rather than new building complexes will provide the site's identity. An urban domain constituted by landscape elements, Tree City attempts to do more by building less, producing density with natural permeability, property development with perennial enrichment.

OMA's plan, titled *Tree City*, consisted in a series of clusters, represented in drawings and models as perfect circles of varying sizes, tied together by a network of informal, curvilinear paths. The round clusters could grow or contract without destroying the logic of the plan – an intelligent strategic choice which highlighted the changing character of the natural environment.[21]

Tree City is an urban project whose method is very distant from traditional urban design and, in fact, it incorporates the lexicon and the cultural position of a landscape project: it uses nature as a primary design material and it rethinks the city as an evolving habitat subject to the passing of time. It is not by chance that it is in the late 1990s that the term landscape urbanism becomes predominant in the architectural discourse and the Downsview Park competition can be considered a key moment for the evolution of the discipline;[22] the finalists all implemented this attitude, in certain cases even more radically than OMA.

Tree City, at least on the surface and in its graphic representation, longs for form, for the return of readability, but deep down celebrates the city as a field of Brownian motion without real control, an effervescent bubbling of abstract clusters which wax and wane with time. There is here a striking contradiction between the clarity and cleanliness of the diagrams and the actual fluidity and potential messiness of the process itself – a process that was deemed so unclear that it never really started, leaving the proposal at the stage of hypothetical plan. Perhaps, then, the reason the project did not go further is to be found in the fact that, beyond the beautifully packaged exterior, the project was actually a piece of realpolitik. Rather than selling a park as a model for the city, Tree City reshaped the park following the ebullient logic of the capitalist city itself.

Tree City applied to a natural environment the same working logic of the blocks of a grid city OMA already celebrated as autonomous islands in many of their projects. Ultimately, the proposal replicated and celebrated the condition we see already in the post-industrial metropolis at large – that is to say a formless carpet of development punctuated by core spaces of growth. In doing so, Downsview's merging of park and city seems to recall one of the most important, though often underestimated, statements of modern architecture: Laugier's 1753 exhortation to look at the city as if it were a forest.[23] As much as the modern city has frequently been described as a machine, this second analogy has long haunted the imagination of architects and administrators alike. If the city as a machine recalls in a very direct way the idea of the urban as a field of production, the idea of the city as a forest is subtler and ideologically ambiguous. At the same time, the city as a forest seems to suggest the contradictory presences of exploitable biological potential on the one hand, as well as of an idyllic, arcadian and untouchable virgin territory on the other. It is perhaps an image that could only arise in an era of colonization, as it recalls the violence of conquest as well as the fascination with the uncontaminated way of living of primitive people that was typical of Laugier's century. Two centuries afterwards, Mies would express a similar sentiment with a radically different word, calling the twentieth-century city 'a jungle'[24] and therefore addressing explicitly the potential for conflict and barbarism that were already implicit in Laugier's statement.

To draw a similitude between the city and a forest, or a jungle, is actually a very fitting way to describe the relationships of power at play in the modern capitalist metropolis. Such an analogy is not meant to address 'nature' per se, but rather to argue against traditional symbolism, and for a city which is likened to an organically growing body. If on the one hand we have a clearly hierarchical, formally defined system which often adopts an anthropomorphic model clearly delineating what is the head and what the limbs, on the other hand we have a homogenous field in which hierarchies are much more layered and therefore complex and difficult to read.

Moreover, the men of the forest are reduced to beings without a history and without a culture; beings whose conflicts are non-ideological, mere struggles for survival. In a word, animals.[25]

To celebrate the city as a natural occurrence, to accept that its inhabitants are animal bodies, means then to accept its political and economic physiology as an unavoidable, 'natural' process. There is a whole genealogy of architectural case studies in which the rhetoric of nature has been very explicitly used to pass off the system of the capitalist city as 'natural' – perhaps, Haussmannian Paris is the clearest example of such strategy, with its thousands of new planted trees which were Haussmann's great pride and which made Paris into a homogeneous, blank background: a city that had to cater to the disparate tastes of 'nomads' – or animals – from all over the world.[26]

From several points of view Downsview seems to represent the climax of this process – on the one hand, it literally proposes seeing the city as a forest, and on the other hand, it sells itself as an economic project before being an architectural one. Downsview represents our final acceptance of the idea that the city is an uninterrupted carpet of productive potential, a carpet marked by the unpredictability that characterizes the natural world, but also the so-called risk society. And it is here that the 'naturalistic' character of Downsview is at its strongest: in its refusal to settle on a fixed shape, in its insistence on growth and development over shape. From a formal point of view, the scheme is marked by three archetypes: the circular shape of the clusters, their archipelago distribution and the underlying phantom of a grid. And unsurprisingly cluster, archipelago and grid are indeed three configurations that are typical of the capitalist city-as-a-forest.

The competition material for Tree City is far more reassuring than that of Rahm and Mosbach's Jade Eco Park. For starters, the schematic idea behind its plan is easier to understand than the more conceptual and rarefied project for Taichung. Moreover, OMA's slow-growing groups of trees feel pleasingly familiar if compared with the environmental manipulations of Rahm's project. The Jade Eco Park devices, with their technological sophistication, promise to deliver something that has have never experienced before,[27] and large audiences which are notoriously conservative cannot but react with scepticism towards this.

However, ultimately, Rahm and Mosbach's proposal, even in its radical realism – the acceptance of the existing condition as polluted and undesirable – is actually far more optimistic than Tree City. In its apparent determinism, the Eco Park leaves in fact open-ended what is perhaps the most important thing: the actual human experience. The architects cannot fully forecast its users' behaviour nor what kind of cultural understanding they'll have of the site – something that on the contrary is masterfully controlled and played with in OMA's case. OMA's narrative is solidly established, its goals precisely inherited from a century-long cultural background, and this is visible from the well-rehearsed socio-economic arguments of the project statement as well as its confident graphic representation; the whole project shows how the rhetoric of nature has become perhaps the last, and most powerful Trojan horse for politicians and architects to gather consensus around an urban plan. However, while the Downsview Park does reclaim a form of agency for architecture at the scale of the city, it still leaves open the question of the agency of the actual citizens. It is in this respect that the Jade Eco Park gives an extraordinary response, challenging the city-as-a-forest paradigm in a radical way.

The city-as-a-forest embodies all the contradictions of our relationship with the state of nature, a condition that is seen both as barbarism but also innocence, productive ground but also wild junkspace: field but also jungle. It is to this jungle that the Jade Eco Park reacts by becoming, quite explicitly, a machine: while the visitor is allowed to become fully just an animal body, architecture takes control of the jungle. But Mies's and Laugier's 'jungle', ultimately, is a narrative trope before anything else: that is to say, a cultural trope, rather than a spatial one. On the contrary, by working on physical reactions, the Jade Eco Park asks the users to recuperate a primordial form of agency: the awareness of one's own body.

In doing so, Rahm and Mosbach's project stands out as it questions the very use of the word *landscape*. Landscape, as a term, is inextricably linked to *sight* – a sense that becomes almost irrelevant in the haptic environment of the Jade Eco Park. And, after all, the highest kind of critical agency lies perhaps in the possibility to rethink and redefine the intellectual categories we work with as architects, and as human beings.

Notes

1 A well-known trope starting with Archizoom's seminal No-Stop City project (1970). Perhaps the sharpest author who has written on the constructed nature of our atmospheric environment is David Gissen; see for reference his 2014 *Manhattan Atmospheres: Architecture, the Interior Environment, and Urban Crisis*. Minneapolis, MN: University of Minnesota Press, as well as his 2012 *Subnature: Architecture's Other Environments*. New York: Princeton University Press.
2 Catherine Mosbach is principal of Mosbach Paysagistes; the distinction between *architecte* and *paysagiste* is quite clear in French culture as the two disciplines are marked by different training, methodology and tradition. Although 'landscape designer' is the literal translation of *paysagiste*, it is important to notice that this is somewhat a misnomer as the term *paysagiste* implies a much more specific competence.
3 The Jade Eco Park is a collaboration between Philippe Rahm Architectes, Mosbach Paysagistes and Ricky Liu & Associates. As the present text is concerned with the conceptual background of the project, we will not discuss Ricky Liu & Associates' contribution as their role – however crucial – is linked to the actual construction of the park. The text is primarily focused on the development at the scale of the park of themes that were already present in Rahm's work and therefore will mostly focus on Rahm's methodology and ambitions.
4 Philippe Rahm Architectes (2014) *Constructed Atmospheres: Architecture as Meteorological Design*. Milan: Postmedia, chapter 3.2.03, unpaginated.
5 As it substituted the site of an industrial-sized slaughterhouse with a leisure and educational centre.
6 As it is made clear in the book on La Villette edited by Marianne Barzilay (1984) *L'Invention du Parc: Parc de la Villette, Paris, Concours International*. Paris: Graphite.
7 1955 interview with Mies, quoted in Detlef Mertins (2001) 'Living in a Jungle: Mies, Organic Architecture and the Art of City Building', in Phyllis Lambert (ed.) *Mies in America*. Montreal: CCA, p. 633.
8 'Agency is described as the ability of the individual to act independently of the constraining structures of society', write editors Nishat Awan, Tatjana Schneider and Jeremy Till (2011) in the introduction to *Spatial Agency: Other Ways of Doing Architecture*. London: Routledge, p. 30.
9 'Human agency and structure are logically implicated with one another': Anthony Giddens (1987) *Social Theory and Modern Sociology*. Cambridge: Polity, p. 220.
10 Philippe Rahm Architectes, *Constructed Atmospheres*, chapter 1.2.02, unpaginated.
11 Aaron Betsky (2008) *Out There: Architecture beyond Building*, volume 1. New York: Rizzoli, p. 135.
12 This is the main thesis put forward by Philippe Rahm (2009) in *Architecture météorologique*. Paris: Archibooks.

13 Kenneth Frampton, most notably, had already put forward this possibility. See for instance, 'Intimations of Tactility: Excerpts from a Fragmentary Polemic', *Content* 12(1) (August 2007); the theme of tactility that Frampton explores in this text had already appeared in his work of the 1980s and 1990s.
14 'Molecular' and 'meteorological' are terms Rahm himself has put forward in the aforementioned *Architecture météorologique*.
15 Philippe Rahm Architectes, *Constructed Atmospheres*, chapter 3.2.03, unpaginated.
16 See for instance, Philippe Rahm (2006) 'Interior Weather', in Giovanna Borasi (ed.) *Environ(ne)ment: Approaches for Tomorrow*. Milan: Skira.
17 Projects such as Rahm's *Public Air* for Copenhagen can be seen indeed as even larger scale applications of the principles at play in the Jade Meteo Park, that is to say architectures aimed at constructing a sensory landscape *beyond sight*. Philippe Rahm Architectes, *Constructed Atmospheres*, chapter 3.2.01, unpaginated.
18 The working relationship between Alphand and Haussmann is discussed in many passages of Georges-Eugène Haussmann (1890–1893) *Mémoires du Baron Haussmann*, 3 vols. Paris: Havard, as well as in a long list of secondary sources starting from the well-known Sigfried Giedion (1982 [1941]) *Time, Space, and Architecture: The Growth of a New Tradition*. Cambridge, MA: Harvard University Press, p. 764.
19 The most complete discussion of the competition entries can be found in Julia Czerniak (ed.) (2001) *CASE: Downsview Park Toronto*. Munich: Prestel.
20 Quoted in Charles Waldheim (2001) 'Park=City? The Downsview Park Competition', *Landscape Architecture Magazine* 91(3), p. 84.
21 Other great assets of the entry were the rhetoric of the text, which underlined how trees would literally become *the* new city, and the graphics curated by Bruce Mau, which, with their uncluttered feel, seemed more accessible than the complex infographics preferred by the other entrants.
22 As rightly underlined by Charles Waldheim (2006) 'Landscape as Urbanism', in Charles Waldheim (ed.) *The Landscape Urbanism Reader*. New York: Princeton Architectural Press, pp. 46–51.
23 Marc-Antoine Laugier (1753) *Essai sur l'architecture*. Paris: Duchesne, p. 259.
24 'There are no cities, in fact, anymore. It goes on like a forest. That is the reason why we cannot have the old cities anymore; that is gone forever, planned city and so on. We should think about the means that we have to live in a jungle, and maybe we do well by that.' From a 1955 interview with Mies, quoted in Mertins, 'Living in a Jungle', p. 633.
25 For a thorough discussion on the philosophical implications of the dialectic between man and animal, see Giorgio Agamben (2004) *The Open: Man and Animal*, trans. Kevin Attell. Stanford, CA: Stanford University Press.
26 Haussmann explicitly calls the citizens of modern Paris 'nomads' in Georges-Eugène Haussmann (1890) *Mémoires du Baron Haussmann*, vol. 2. Paris: Havard, p. 177.
27 As of 2016, the park is close to completion and the devices do perform as imagined in the competition material.

6
PLANETARY AESTHETICS

Peg Rawes

This chapter examines how Agnes Denes's 1970s map projections and Buckminster Fuller's energy slave maps (1940–1972) can be understood within a contemporary biopolitical discussion about aesthetics and ecology. I suggest that these architectural and artistic planetary visualizations preview current aesthetic and ecological preoccupations with spatial, social and biological understandings of 'life' in the humanities, architecture and the visual arts. In particular, the chapter explores how a biopolitical explanation highlights 'ratios' of ecological and economic information in the artist's and architect's practices. In addition, I suggest that Denes and Fuller's work has a renewed historical valency, given the current urgency for addressing climate change on a planetary scale.

The stakes of ecology's and economics' roots in *oikos* are therefore even greater than when discussions of energy conservation were underway in the 1940s and 1950s USA, in which Fuller energetically took part,[1] or the political environmental movements in the late 1960s, with which Denes's work has resonance. These mid-twentieth-century environmental contexts help to highlight the historical formation of present-day research that investigates the critical and poetic formations of human and non-human ecologies.[2] In addition, their fascinating 'forecasts' are reappraised by being situated in relation to questions of 'data', 'information' and rationalism, and by research into the aesthetics of data visualization that examines how biology and politics operate representationally.

As such, Denes's and Fuller's maps have biopolitical value because they show how modern forms of rational thought about the planet organize society (especially geometric forms), and how art and architecture visualize these. In addition, these planetary images reveal aesthetic forms of imaging data and biological 'information', which now intersect with current debates about the role of information and data in climate change geopolitics, and in relation to the visibility or invisibility of non-global communities; for example, think about how the 2015 Paris COP21 debates focused on agreeing a 1.5 per cent temperature increase, which was resisted by many of the

BRIC nations, yet seen as essential to ensuring any kind of environmental well-being of smaller coastal communities, including the Maldives.[3] In addition, because modern-day climate science is inherently concerned with its *powers* of prediction and forecast, these 'proper' scientific principles are then further open to ideological contest by different interest groups, including those professionally involved in 'agnotology' (climate change deniers), such as political lobbyists for the oil and mining industries.[4]

Of course, Denes and Fuller are practitioners whose visual vocabulary and writings precedes our present-day Anthropocene terminology to describe the societal manipulation of planetary resources through carbon technologies. However, both practitioners knowingly used cultural and planetary information about man-made environments and earthly resources in these images; Fuller's concern with improving the distribution and efficacy of energy technologies in the post-war period, and Denes's focus on ecological approaches to agriculture, farming and, in these projections, playful references to foodstuffs (up to 2015 her practice continued to focus on these concerns[5]). So, despite their projections being produced before present-day Anthropocene discourse, the images are striking previews of globally distributed human and non-human interactions, and the biopolitical ratios that compose visual forms of environmental information.

This is especially relevant given how aesthetic and rational modes of visualization may be used intentionally to unsettle stable bodies of knowledge or meaning, positively but also in disturbing ways, especially when representing the differentiation of human and non-human 'life', and the rights that are given to them. Denes's and Fuller's engagement with organic and inorganic figures (e.g. an invertebrate, a foodstuff, a geometric object, a continent or a human figure) reflect the way in which corporations and governments now look for personal forms of data that can be used to 'design' global 'big data' markets, such as the healthcare and education sectors (e.g. biometric data about our well-being, happiness or health). Denes's and Fuller's cartographies are therefore also situated within a broader context of the aesthetics of technologies for visualizing the economic values of 'data' for a society. Moreover, as I argue below, and has been shown by other researchers, Fuller was actively involved in these ideological practices in the post-war period.[6]

Recent biopolitical discourse has sought to demystify the belief that the distribution of data (especially at a societal, technological and governmental level), naturally brings with it positive democratic values for all. Michel Foucault's 1970s study of neoliberal forms of governance and its territories has helped researchers unpack the affects and effects of macro-scaled management of human life as well as the formation of the individual at micro and biological scales of organization.[7] In particular, Foucault's theory shows how political organizations of biological life produce negative rationalist forms of individualization. Following Foucault, Giorgio Agamben has argued that the most negative biological classification of 'life' is *zoe*, what he calls 'bare life', rather than the 'legal' definition of life as *bios*.[8] *Zoe* designates a form of life that is exempted from society, 'other' or non-human. Such subjectivities are excluded from 'legitimate' legal, cultural and environmental value systems. In its most violent enunciation, *zoe* is the non-human subjectivity who is formed by systematic governmental technologies of dehumanization.

Hence, biopolitical studies offer architectural and arts researchers the capacity to examine how the techno-scientific and governmental organization of society – and, consequently, local interactions with our environments – are not neutral or natural processes of distribution. In addition, it exposes how certain kinds of subjectivity and of life are exempted from these rational systems of design (or 'control'). Moreover, discussions about what constitutes non-human life rights (which can extend from the rights of indigenous or aboriginal populations, to absolutely 'non-human' wildlife populations, such as polar bears, penguin colonies or Amazonian canopies) intersects with architectural and artistic practices which strive to design against the impact of climate change and threats to societal and environmental well-being. Such alliances are considered increasingly necessary in the face of poor environmental governance that favours short-term economic gain, unethical depletion of resources, and the concomitant dehumanization of communities who resist economic development.[9]

Denes and Fuller therefore belong to an earlier generation of practitioners who explored rational forms of visualization, especially scientific knowledge, but they are now valuable for being prescient versions of the dysfunctional and complex forms of *oikos* that concern us today. Thus, while we might agree that Denes's projects are aesthetically beautiful, these 'absurd' images disjoint notions of 'natural' planetary and human organization. The maps highlight the capacity for 'data' visualizations to reveal dysfunctional global distributions of material, organic and environmental resources, especially given current humanitarian issues of access to food, agriculture or the impact of war on vulnerable regions. Buckminster Fuller's energy slave maps take the term 'slavery' to promote a 'universal' principle of machinic labour for reducing inequality in global labour. His coupling together a technocratic approach to improving carbon energy efficiency with a historical term for the dehumanization of 'others' (especially of black and minority populations), is now even more disturbing given Agamben's critique of state violence against individuals and communities deemed 'outside the law' and, most recently, given high rates of exploitation of migrants in global agricultural and architectural construction sites. Thus, while they produce quite distinct versions of 'non-human' alterity – one aesthetically inventive, one utopic, yet troubling – Agnes Denes's and Buckminster Fuller's projections precede current practitioners' critical engagement in the visual, aesthetic and techno-scientific formation of ecology insofar as each represents planetary ratios of alterity, distortion and 'otherness'.

Alien projections

My discussion begins with reference to an image that marks a historical shift in the visualization of the planet: NASA's 1968 'Earthrise' photograph, the 'first' image of the planet from Apollo 8. As has been well documented 'Earthrise' locates a shift in understandings of human and planetary relations from a pre-astronautical projection into a technological form of evidence. Imaginary and predictive visualizations of the world are corroborated and the planet's aesthetic value is refreshed in the photographic representation. Again, as has been well-covered by researchers, in the same year a social and political environmental shift takes place in the publication of Stuart Brand's *Whole Earth Catalogue*, which emphasized the utopic rights of the modern individual and of a post-war neoliberal environmental imagination.[10] Fuller's cartographic energy forecasting and Denes's alien projections are situated on either side of these 1968 human–planetary revisionings; each a re-imagining of the relationship between scientific and ecological definitions of life. Each providing a visualization of the rational and representational 'ratios' that compose natural and human-made ecologies, still resonant with current-day environmental and climate-change concerns about the existence (or extinction) of human-to-nature and non-human relations.

In the decade after 'Earthrise' Denes produced a series of ecological and geometric maps titled 'Isometric Systems in Isotropic Space: Map Projections' (1974–1976). Interestingly, Denes also refers to her work, including the projections, as 'alien'.[11] Each drawing presents the Earth as an isometric 'object', including shapes which are easily identifiable as geometric objects, such as pyramids, cubes and dodecahedrons. But this series of 'worlds' also includes isometric 'translations', including toroids and ovoids, which classify the Earth as less exclusively 'mathematical' objects. Also, for these 'other' planets, Denes uses names of organic objects – snail, egg, lemon, doughnut or hot dog – rather than 'proper' geometric names, so that each map

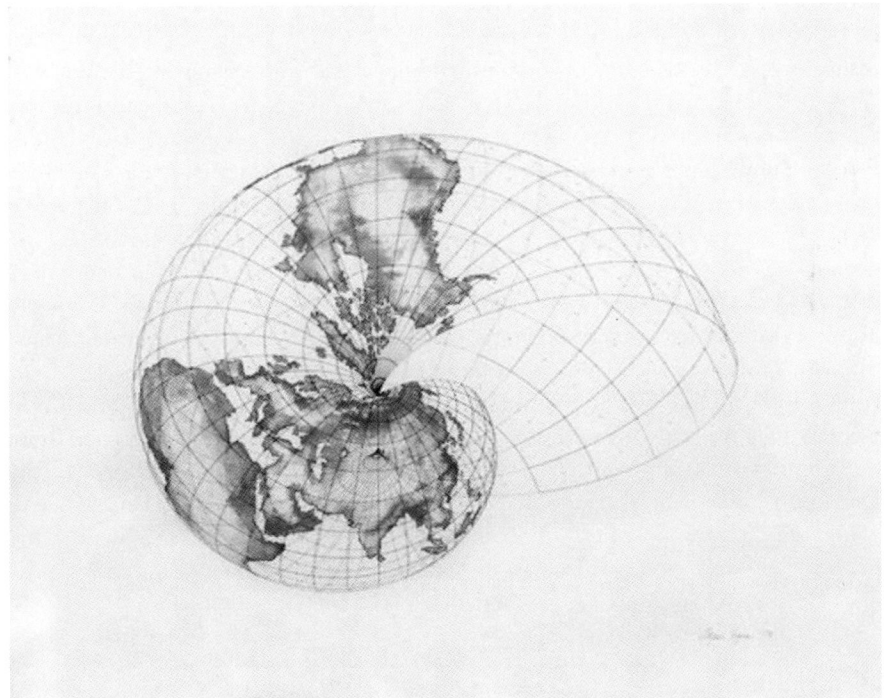

FIGURE 6.1 Agnes Denes, Isometric Systems in Isotropic Space Map Projections: The Snail, 1979. Ink and gouache on paper and Mylar. 24 × 30 inches. Credit: Copyright Agnes Denes, Courtesy Leslie Tonkonow Artworks + Projects, New York.

oscillates between an identifiably mathematic object and an 'other' register: everyday global foodstuffs and objects from nature. Properly 'scientific' or technical drawings are simultaneously projected into an aesthetic mode of understanding (an aesthetics of taste, literally) in which the precise rationale of mathematical thought and objects shifts from corroborating scientific reasoning into an aesthetic reimagining of the world. The Earth becomes remade as a series of denatured projections that bring together the everyday human world of foodstuffs and natural forms at a planetary scale.

Denes's practice is therefore composed of aesthetic operations that visualize political, social, technological and environmental modes of planetary relations (i.e. ratios). Her geometric 'elements' are knowingly 'absurd' earthly idealizations, rather than pure platonic forms. The axiomatic power of geometric reasoning is redirected to show planetary ecologies of material and spatial natures, resulting from an aesthetic practice that also relates to architects and artists today who critique the over-consumption, pollution and resource-depletion of our human and non-human environments.

The maps are also extremely beautiful. But what is significant about them is the material power of disruption which is given to the 'othering' of geometric objects. By transposing an entirely commonplace, yet absurd, nomenclature, rational definitions of human, animal and planetary relations are destabilized. In addition, while these alien projections do not explicitly articulate a critique against the economics and *oikos* of global food trade that Denes undertakes in *Wheatfield: A Confrontation* in 1982, they do register her concern with the rational, economic and ecological formation of *oikos*. *Wheatfield*'s 'amber field' locates the artist in front of the Twin Towers. Still an iconic and globally disseminated image of 'ecological thinking', this artwork captured the imbrication of built, social and environmental relations that still underpin twenty-first-century urbanism. At the time, Denes said the project questioned the 'mismanagement, waste, world hunger and ecological concerns' which constitute modern regimes of 'food, energy, commerce, world trade, economics'.[12] Ten years earlier, although not directly addressing the economics of food production, Denes's planetary defamiliarizations raise questions about the imbrication of rational scientific knowledge (e.g. NASA's newly confirmed scientific visualization of the planet) with the globalization of food industries.

Preceding *Wheatfield*'s agricultural intervention, literally under the World Trade Center, Denes's maps of inventive classifications of human and planetary relations represent a poetic critique about the anthropocentric rationalization of the Earth into human-focused resources. More explicitly visualized ten years later in *Wheatfield*'s invocation of global economics, agriculture and food production, the maps suggest an 'eco-logical' form of reasoning, where aesthetics is used to produce a disruptive rationalist logic, perhaps even a 'humane' rationalism.[13]

Non-human ratios

Buckminster Fuller's utopian design practice during the 1940s, 1950s and 1960s is deeply concerned with the translation of biological, material and political information into metrics and data visualization, and which therefore resonates with contemporary biopolitical discourses of human and non-human relations. Central

Planetary aesthetics **83**

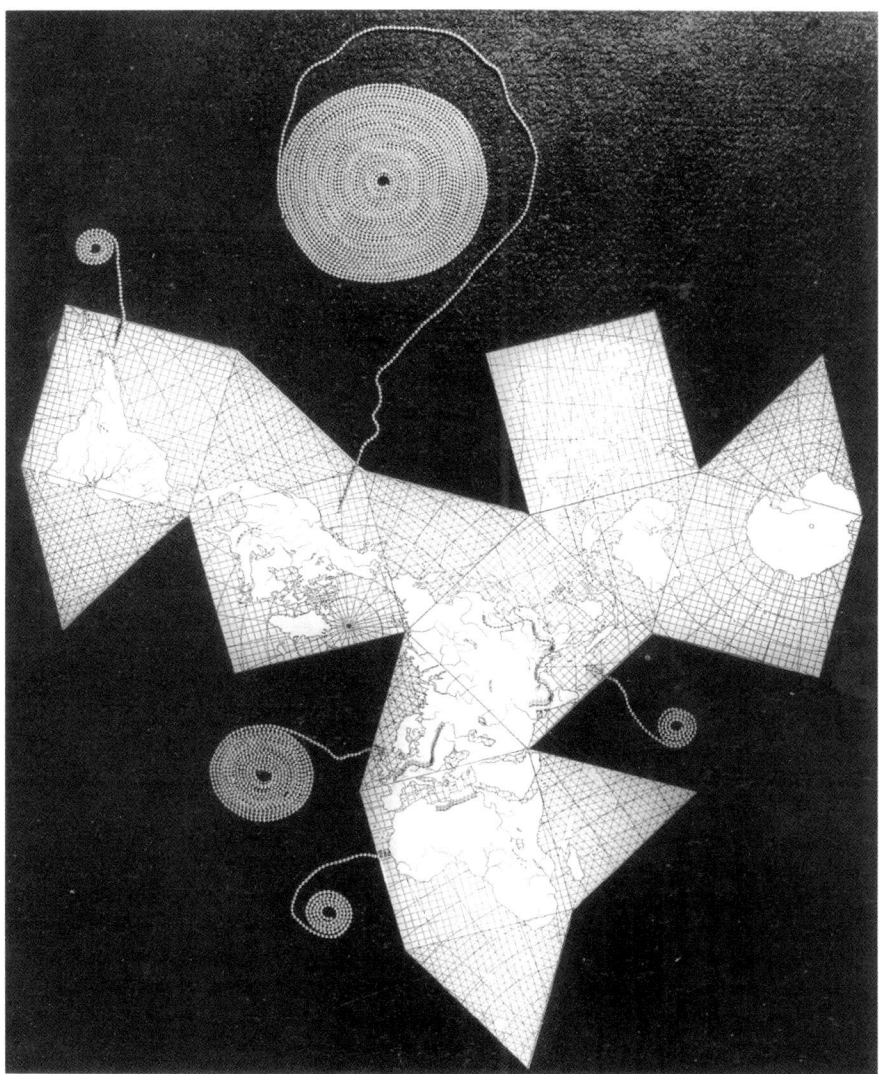

FIGURE 6.2 Dymaxion map of energy slaves, 1945. Courtesy, The Estate of R. Buckminster Fuller.

to his post-war practice was the development of the Dymaxion maps which visualized the planet's entire landmass in a single 'unified' island within one ocean. Fuller argued that this rational geographic re-envisioning of the Earth's continents and seas succeeded in reducing the distortion of their respective shapes and sizes. Dymaxion maps could 'project' more efficient technological solutions for managing global energy resources and, consequently, enable more effective social and political scenarios to be planned. Also, interestingly, Denes herself spoke of the visual resonance between Fuller's and her maps when her book of projections was published in 1979:

> Map Projections takes our globe and transforms it into several trigonometric shapes, while keeping correct mapping measurements. When Buckminster Fuller whom I met a few years before his death saw the book he exclaimed: 'I should have done this!' He was so impressed, he sent me his dymaxion map as a present. Map Projections distorts perspective and creates these absolutely funny forms from our globe, such as a Doughnut (tangent torus), where the North Pole and the South Pole meet in the hole of the Doughnut pulling in the continents; Snail (helical toroid), the Egg (sinusoidal ovoid), The Lemon (prolate ovoid), and a Hot Dog, a Pyramid, a Dodecahedron and a geoid, which allows a continent to escape the earth and form the moon. The longitude and latitude lines are unraveled and the continents are allowed to float in space and assume new configurations. Mathematical forms are projected over fluid space to create maps that are witty and mysterious. As soon as a form was created, I dissected it into its fragments to yield further beauty in a new form. Playing such games with our orbiting home, our only home in the universe is brazen, but makes for good art.[14]

'Dymaxion' was coined by Waldo Warren, an advertising specialist for the Marshall Field Organization, in 1929. It synthesizes the words 'dynamism', 'maximum' and 'ions' and is Fuller's patented term for his 4D 'Energetic Geometry'[15] projects in the decade before World War II including his 'house of the future'.[16] Ten years later, after the outbreak of war, Fuller deployed the term to describe another design product, a new map of the world, which was intended to correct the spatial distortion and disjunction of the continents that are retained in the Mercator and Robinson projections. Fuller argued that both existing projections inflated the scale of Greenland, and the Mercator projection also reduced the proportion of Antarctic landmass to a 'strip along the bottom'. For Fuller, both projections failed his utopic desire for a consistently geometric map of global information, thereby reinforcing existing geopolitical disjunctions and inequalities; the 'inherently disassociated, remote, self-interestedly preoccupied with the political concept of its got to be you or me; there is not enough for both'.[17] Instead, by bringing together all regions and all seas into one – 'one great continental archipelago lying within a one world ocean'[18] – Fuller set out to visualize 'a precise means for seeing the world from the dynamic and comprehensive viewpoint'.[19] Rather than repeating the mistakes of Mercator and Robinson, which reproduced a distribution of planetary 'information' through regionally differentiated landmasses (and by implication, of different socio-political ideologies), Fuller saw the Dymaxion map as a means to generating 'global information' with 'negligible distortion' that would aid strategic geopolitical resources and planning, including ballistics and automated aircraft.[20]

In a more explicitly environmental section of his commentary, Robert Marks argues that Fuller's dynamic geometric forms of energy efficiency also represent a kind of ecological thinking; for example, Fuller's interest in the organization of information, including environmental and organic resources and population, represents 'a concept of major and minor ecological patterning, that is, regularities in the relations of organisms to their physical environment. . . . birds' seasonal, world-sweeping migrations represented to Fuller a major ecological patterning; birds nest-building and "local regenerative to-and-fro-ing", what he regarded as the related minor ecological patterning'.[21] This definition of Fuller's ecological approach is still recognizable in architectural design studios today, especially for the professionals who use digital software and computational systems to organize material, social and spatial information. However, until recently, architectural design has tended not to consider these designs within a discussion of biopolitics, or of human and non-human relations.[22]

Fuller's universal mappings of human and non-human energy clearly are inherently biopolitical because they visualize the utopian and rational belief in the capacity of techno-scientific methods for managing land and energy use. These principles are particularly strongly represented in the energy slave maps which Fuller developed during the 1940s, and first published in February 1940 in *Fortune* magazine under the title 'The World Energy Map'. Within the context of acute global concerns about the war, energy distribution and resources during this time, Fuller's energy slave maps are evidence of a biopolitical form of planetary aesthetics, especially as the maps represent the unequal distribution of human and non-human energy sources as rational units of 'global information'.

Reproduced in *The Dymaxion World*, the World Energy Map shows in 1950s USA a ratio of 347 non-human energy-producing units to every human unit (i.e. 'family'). Statistical data appending this visualization translates the energy required by the USA into 2,774 per cent of the total global mechanized energy. In contrast, South America is attributed with 114 per cent of the world's energy slave allocation, 646 per cent to Europe, 152 per cent to Africa and the Mediterranean, and zero in Central America.[23] Of course, while Fuller uses the notion of 'slave' as a cypher for 'robotic' or mechanized energy capacity, in the context of today's issues of migrant labour forces and the inhumane global trafficking of slave labour, this language is challenging and aesthetically dissonant (aesthetics here meaning the *displeasurable* agitation felt in attempting to 'grasp' the non-human power of nature: for Kant, this was the Sublime; in modern theories of biopower, this agitation is not so much a fear of the 'natural', but the dehumanization of the 'other' – the migrant, the refugee, the slave).

Thus, while Fuller intended these maps to be objective rational explanations of projected energy needs, they are actually a troubling biopolitical planetary aesthetic in which projection or forecasting is aligned with a utopic belief in technological management of resources. This is underscored in Fuller's annotation on the following page for a 'graph of the rate of attainment of world industrialization to 1952 and the projected rate to the year 2000'. The visualization predicts that a global implementation of energy slaves into the post-war crisis in energy efficiency and supply will reduce the resulting gap between the 'haves' and 'have nots' after 1972. However, his annotation ends far less confidently about reaching global energy equity – presumably updated by Fuller for re-publication in 1973 – in its prediction that the tendency for 'incitable' socialist revolution in the 'have nots' also increases after 1972, because of the even greater inequality of resources available to the 'more unduly privileged minority of the "haves"'.[24]

Fuller's attempt to produce an entirely rational order of global 'information' for addressing industrialized society's energy requirements is ultimately determined by his ideological belief in technological and scientific projection. Thus, even if he wished to redress the violent history of slavery against the black population (as Marks suggests), and to invent a positive techno-scientific representation of non-human labour, his energy slave maps remain problematic, both as reminders of histories of slavery, and because they are so prescient of current inhumane global distributions of human labour.

Fuller's utopian analysis of global human power relations contrast with Donna Haraway's critique of techno-scientific subjectivity, and contemporary new materialist feminist thought which questions normative humanist/universalist histories of ideas. Haraway's biopolitical 'critical sympathy' is elegantly expressed in *Modest Witness*, written in 1997, but still exceptionally futural for tackling today's issues of resource and environmental inequality. Here Haraway distinguishes between an ethical non-humanist subjectivity and dominant 'neutral' techno-scientific epistemologies and universal subjectivities:

> Shaped as an insider and an outsider to the hegemonic power and discourses of my European and North American legacies, I remember that anti-Semitism and misogyny intensified in the Renaissance and the Scientific Revolution of early modern Europe, that racism and colonialism flourished in the travelling habits of the cosmopolitan Enlightenment and that the intensified misery of billions of men and women seems organically rooted in the freedoms of transnational capitalism and techno-science. But I also remember the dreams and achievements of contingent freedoms, situated knowledges and relief of suffering that are inextricable from this contaminated triple historical heritage. I remain a child of the Scientific Revolution, the Enlightenment and techno-science.[25]

Whereas Fuller endorses a positivist form of humanist universalism, Haraway's analysis of power focuses upon rethinking the ethical relationship between human and non-human modes of life (i.e. animal and technological) as foundational to our capacity to improve ratios of environmental equality, well-being and survival today. For Fuller, this is the efficient distribution of energetic powers, human and non-human. For Haraway, the challenge is to fully enable the positive *alterity* of non-human techno-social relations. Both are concerned with the technological, material and ecological practices that constitute planetary relations and resources, but where Fuller's interest is determined by economic and energetic efficiency, Haraway wishes to restructure the relationship between *oikos*, human life, technology and economics, to realize more ethical forms of social and political biopower. Fuller's belief in the capacity for society and its resources to be improved through a rational techno-scientific management systems is therefore substantially distinct from the ethical, that is 'modest', critique of these universals proposed by Haraway: and which, in this sense, resonates more closely with Denes's promotion of 'alien' distortions. Produced thirty years after Fuller's Dymaxion map, Denes's alien maps serve to remind us of the dysfunctional rationalism of global food trade and the uncritical belief in an anthropogenic control of the environment: thus the 'hot dog' and doughnut represent a kind of sympathetic but critical planetary aesthetic, and are closer to Haraway's 'modest witness'.

Denes's maps are 'alien' rational imaginaries of the Earth and of its environmental resources. In contrast to these playful visualizations, Fuller's utopic forecasting of mechanized energy distribution uses the deeply uncomfortable language of slavery, re-visualizing this dehumanizing history in a global techno-scientific aesthetic: images that are still factually and aesthetically disturbing. Denes and Fuller also come under a biopolitical lens because of cartography's historical legacy of rational thought. Different modes of rationalism are generated: both practitioners' images operate from within logics of scientific visualization; both use a projective, measured and mathematical logic of representation. Spatial, geometric and geographical ratios are manipulated in order to propose new social, material and power relations. In Denes's map projections, geometric and mathematic visualizations are deformed. Standard projections of the ratios between landmasses and sea are made strange. The 'facticity' of the Earth is de-rationalized by its transposition into isotropic geometric forms. A further denaturing is emphasized by the application of 'absurd' yet familiar names of foods. On two levels then, Denes produces a dysfunctional geometric rationalization of the Earth and its resources: first by defamiliarizing its spherical geometry, and second, by applying non-scalar organic identities to each of the forms. Fuller's rationalism is much more troubling; more explicitly comprising a biopolitical project because of Fuller's active engagement in visualizing systematic techno-scientific management of human and non-human life. Fuller's deployment of the term 'slavery', represented through the silhouette figures manifests this ambition most strongly. Global relations are biologically and politically re-envisioned through a 'corrective' reimagining of the violent history of dehumanizing 'non-human' lives into a utopic vision where mechanized labour is forecast to successfully meet our energy needs.

For historians and theorists of ecological practice, these architectural and artistic visualizations show the imbrication of the visual politics of ecological and planetary discourses and the effects of rational thinking. Foucault's analysis of how particularly inhumane forms of rational thought were systematically developed in seventeenth-century governmental regimes helps to critique the relationship between rational and aesthetic accounts of social, spatial and environmental interactions.[26] Given that these neoliberal logics clearly still manage the distribution of economic, societal and environmental resources on a planetary scale, the need to examine these dysfunctional, distorted and unequal ratios remains a pressing concern for effective critical ecological inquiry and practice.

Notes

1 See Daniel Barber's (2013) discussion of the post-war energy consumption and architectural design. Also see Braham and Willis (2015).
2 See, for example, Rawes et al. (2016).
3 See, for example, Donner (2015).
4 See Lorraine Code (2013, 76) on 'agnotology'.
5 Denes is still active in developing projects, for example, *Living Pyramid*, a Rockefeller-funded public art project in Socrates Park, New York (2015), and the 2015 re-making of *Wheatfield* in Milan. See Leslie Tonkonow Artworks + Projects gallery, www.tonkonow.com/ (accessed 1 September 2015).
6 See Barber (2013).
7 See Foucault (2008). Also see his 1984 seminar in which he analyses a seventeenth-century French essay about the 'well-governed state' to identify inhumane forms of rational thought which govern and design society through a system of 'boards', including the management of the built environment.
8 Agamben (1998).
9 See Rawes et al. (2016).
10 See, for example, Scott (2016).
11 Denes (1993, 387).
12 Ibid., 389.
13 It is also worth noting that, again in 1968, Denes used the term 'eco-logic' (a term that Guattari then also deployed in his 1989 analysis of the economic, environmental and social formation of global relations in *The Three Ecologies*).
14 Denes (1979) 'Artist statement'.
15 Fuller and Marks (1973, 20).
16 Ibid., 50.
17 Buckminster Fuller Institute, https://bfi.org/about-fuller/big-ideas/dymaxion-world/dymaxion-map (accessed 1 September 2015).
18 Fuller and Marks (1973, 51).
19 Buckminster Fuller Institute website.
20 Fuller and Marks (1973, 51).
21 Ibid., 16.
22 For architectural researchers who do deal with the politics of digital, environmental and ecological information for architecture see, for example, Barber (2013), Scott (2016) and the anthology, *The Politics of Parametricism*, edited by Matthew Poole and Manuel Shvartzberg (2014), including chapters by Rheinhold Martin, Benjamin Bratton, Laura Kurgan, Peggy Deamer and myself.
23 Fuller and Marks (1973, 154).
24 Ibid., 155.
25 Haraway (1997, 3).
26 Martin et al. (1988, 16–19).

References

Agamben, G. (1998) *Homo Sacer: Sovereign Power and Bare Life*, trans. Daniel Heller-Roazen. Palo Alto, CA: Stanford University Press.
Barber, D. (2013) 'Hubbert's Peak, Eneropa and the Visualization of Renewable Energy', *Places Journal*, 20 May, https://placesjournal.org/ (accessed 20 December 2015).
Braham, W. and Willis, D. (eds) (2015) *Architecture and Energy: Performance and Style*. New York and London: Routledge.
Buckminster Fuller Institute (n.d.) https://bfi.org/about-fuller/big-ideas/dymaxion-world/dymaxion-map (accessed 1 September 2015).

Code, L. (2013) '"Manufactured Uncertainty": Epistemologies of Master and the Ecological Imaginary', in P. Rawes (ed.) *Relational Architectural Ecologies*. London and New York: Routledge.

Denes, A. (1979) *Isometric Systems in Isotropic Space: Map Projections (from the Study of Distortions Series, 1973–1979)*. Rochester, NY: Visual Studies Workshop Press and New York State Council on the Arts. Available at: www.artistsbooksonline.org/works/mpjs.xml. (accessed 15 December 2015); Leslie Tonkonow Artworks + Projects: www.tonkonow.com/.

Denes, A. (1993) 'Notes on Eco-logic: Environmental Artwork, Visual Philosophy and Global Perspective', in special issue: 'Art and Social Consciousness', *Leonardo* 26(5), pp. 387–395.

Donner, S. (2015) 'Why We Need the Next-to-impossible 1.5°C Temperature Target', *The Guardian*, 30 December, www.theguardian.com/environment/climate-consensus-97-per-cent/2015/dec/30/why-we-need-the-next-to-impossible-15c-temperature-target (accessed 5 January 2016).

Foucault, M. (2008) *The Birth of Biopolitics: Lectures at the Collège de France, 1978–9*, trans. Graham Burchell. Basingstoke: Palgrave Macmillan.

Fuller, R.B. and Marks, R. (1973) *The Dymaxion World of Buckminster Fuller*. New York: Anchor Press and Doubleday Books.

Guattari, F. (2000) *The Three Ecologies*, trans. Ian Pindar and Paul Sutton. London: Athlone Press.

Haraway, D. (1997) *Modest Witness@Second_Millennium. FemaleMan_Meets_OncoMouse: Feminism and Technoscience*. New York and London: Routledge.

Martin, L., Gutman, H. and Hutton, P. (eds) (1988) *Technologies of the Self: A Seminar with Michel Foucault*. Amherst, MA: University of Massachusetts Press.

Poole, M. and Shvartzberg, M. (eds) (2015) *The Politics of Parametricism: Digital Technologies in Architecture*. London: Bloomsbury Academic.

Rawes, P. (ed.) (2013) *Relational Architectural Ecologies: Architecture, Nature and Subjectivity*. London and New York: Routledge.

Rawes, P., Mathews, T. and Loo, S. (eds) (2016) *Poetic Biopolitics: Relational Practices in Architecture and the Arts*. London: IB Tauris.

Scott, F.D. (2016) *Outlaw Territories: Environments of Insecurity/Architectures of Counterinsurgency*. Cambridge, MA: MIT Press.

7
THE CLOSED LANDSCAPES OF SVERDLOVSK-44 AND KRASNOYARSK-26

Katya Larina

Soviet secret cities, built during the Cold War as a network of high-security centres for research and production, have remained hidden from any map for five decades. These cities were typically enclosed by concrete walls and barbed wire and were known only by the code number that followed their 'ZATO' designation ('Closed Territorial Formation'). The first ZATOs were instituted following the launch of the nuclear weapons programme in the 1950s by the Soviet government. The typical ZATO city contains a population of between 100,000 and 200,000 people. These settlements, based around research and industrial plants, were built as secret and secure places for elite Soviet scientists to work on nuclear weapons development and later nuclear energy research and production. Over time the purpose of the ZATOs became yet more diversified. Today their functions include space research and production and conventional arms engineering, and they may also serve as military ports and bases.

Initially, ZATOs very often were connected to larger transport and energy infrastructure systems, the network of gulags (Vytuleva, 2012, 5) and resource development centres, but despite their networked reality, they remained invisible 'shadow cities', impossible to locate on any map or in any official records until after the collapse of the Soviet Union in 1991. Today, more than forty of these 'closed cities' remain, scattered over the territory of contemporary Russia and retaining a combined population of more than one million. Since the collapse of the Soviet Union, the ZATO functions have changed from purely military and secret technologies to encompass innovation in energy, biotechnology and communication. While many such functions might still require security, there is no longer a need to maintain such extreme secrecy. All the same, even now high boundary walls confine the ZATOs and access to them is highly restricted. In the volatile economic and political circumstances of the post-Soviet era, many ZATOs, long dependent upon highly centralized and government controlled administration, have been left without such rigorous central control, and also their funding has greatly diminished. This has resulted in a significant loss of population and a period of stagnation in these cities that had previously flourished (although secretly) throughout the Cold War. The future of these cities is uncertain – some have been declassified and some have lost their strategic functions. The decline in state demands upon the military industry has directly affected research and production within the ZATOs, thus threatening their privileged status and causing them to seek out ways of operating more independently.

I was involved in several multidisciplinary projects to create urban social economic strategies for these cities, as part of a team of economists, sociologists and futurologists. We were looking for a way to 'open' the closed cities. Our findings indicated that what the process required was a measure of support alongside what was already happening through official neglect: a significant reduction of the top-down control of the state and the identification and amplification of internal and autonomous agency to support quality of life and continued production within these 'shadow cities'. The closed cities actually have great potential for future development. As enclaves for elites, they were provided with high-quality infrastructure and rich urban landscapes. In Soviet times they were prestigious cities in which to live and work, albeit that this prestige was limited to the range of a select and clandestine cognoscenti. As scientific centres, ZATOs have a greater percentage of residents with higher education qualifications as compared to other Russian cities. Despite the intellectual credentials of these cities, we found ourselves facing an absence of initiative in the population for driving innovation and entrepreneurial activity. Individual agency had been dulled by a history of central state control.

What my work sought to undertake was to transform the historic command and control structure of the society through Jane Jacobs's classic formulation of the city as an 'open system', reinterpreted in the work of Richard Sennett into the concept of the open city (2006). Sennett claimed that 'We need to apply ideas about open systems currently animating the sciences to animate our understanding of the city' (2006, 1). The open system, in Sennett's formulation, is a system 'reactive to its environment' (2016), it is constantly evolving by interacting with its changing external conditions. Contrary to this, the closed cities operated in a closed system, which existed in a stable state in a larger environment that was also fixed and artificially maintained, a city where you 'know everything in advance, when you don't learn in a process of building a city' (2014).

Sennett writes: 'Closed means over-determined, balanced, integrated, linear. Open means incomplete, errant, conflictual, non-linear' (2006, 14). The city acts as an 'open system' when it has the internal agency necessary to provide the capability of adjusting to adapt to dynamically changing social-economic, political or environmental conditions. Sennett challenges artificially balanced closed systems, saying that however 'in an open system, the city is to a degree incoherent. Dissonance marks the open way of life more than coherence, yet it is a dissonance for which people take ownership' (2006, 1).

A big part of our work thus comprised the study of the larger, physical environment of the 'closed city', alongside the creation of socioeconomic strategies and policies. The urban landscapes of the ZATOs were, in essence, the spatialization of state-centric and bureaucratic fixity, and thus it became of paramount importance to examine the urban fabric and its relationship to the wider landscape beyond the walls and barbed wire to seek to create an environment in which productive socioeconomic and research activity could thrive. Further to this, however, was the consideration that the landscapes created by central state functions were not either completely dysfunctional, nor were they unloved by their inhabitants. Thus the transformation of the ZATO landscapes would have to take this into account and not try to completely reconfigure the spaces. The landscapes have a unique character that evolved behind the walls in the political epochs of both Soviet and post-Soviet Russia that is expressive of this history, and we did not wish to deny or erase these qualities. Two specific closed cities are examined in this chapter: Sverdlovsk – ZATO number 44 and Krasnoyarsk – ZATO number 26. The analysis of these cities reveals the material properties of the 'closed' urban landscape and the ways in which this differs from that of the more conventional 'open' city.

The urban landscapes of Sverdlovsk-44 and Krasnoyarsk-26 exhibit forms which are typical of the closed cities and which developed primarily during the period in which the cities were completely sealed and secret. First, the cities' structures are populated with fragments of reified images of ideal cities. These fragments, in the form of Stalin, Khrushchev or Brezhnev-era districts have appeared in the city incrementally, due to the successive ideological paradigms (and thus shifting ideal forms) of the Soviet and post-Soviet political context. Second, the structure of the city including the network of public spaces reveals the nature of its urban landscape as primarily an infrastructure for a state industrial machine, controlled, as mentioned above, from the top down. Third, the post-Soviet closed city is now taking on expressions that show the beginning of the process of 'opening up', and these nascent forms are of great interest and show potential for the ZATO to transform into the informal 'open city'. Working in ZATOs I paid special attention to those public spaces with some signs of locally emerging commercial and public activity, places like markets, gateways to the city, railway stations, etc. Addressing the question of the relationships and tensions between the ideas of the 'closed' and the 'open' city, and the strategies and design decisions that may drive 'top-down' and 'bottom-up' operations, I explore the notion of how the 'public' as an ideal may be redefined and transformed in relation to the environment and design of the urban landscape.

Ideal 'cities in the city'

Sverdlovsk-44 and Krasnoyarsk-26 were two of a total of ten 'atomic' zone cities, built solely to further the Soviet nuclear program. As in other ZATOs, they were built originally in the 1950s and 1960s and brought elite scientists, engineers and technicians together in one place to work and live in total dedication to this singular cause. Though it was a privilege to be invited to dwell in these cities, it was also necessary to provide a palliative to the problem of living secretly, thus they were built as walled utopias, highly maintained and unusually well supplied.

They were designed in the svelte Soviet modernist urban tradition, combining the features of the rich ensembles of public squares and promenades, lavish parks and a highly developed social and civic infrastructure. 'Atomic' cities were also traditionally built on stable bedrock in order to minimize the potential ecological catastrophe in the event of a leakage of toxic substances. In Soviet times all ZATOs were listed under a combination of code names and numbers, sometimes with the name of the closest administrative centre: Sverdlovsk-44, for example, is ZATO city number 44, and it lies near to Sverdlovsk. In 1994 during the early days of the new post-Soviet Russia, by order of the Council of Ministers, the presence of 'coded' cities was declassified and Sverdlovsk-44 gained its old informal name – Novouralsk – and Krasnoyarsk-26 became Zheleznogorsk.

Sverdlovsk-44, now known as Novouralsk, is situated in the eastern foothills of the Ural Mountains, a picturesque landscape of mountains, hills, lakes, rivers and wild forest, and there are expansive views of the wider landscape from the high points of the city. In 1946, in accordance with the 'nuclear shield' program of the Soviet Union, the party started the process of building this new, secret, industrial complex from which the highly enriched uranium used for the first Russian nuclear bomb was produced (CNCP, n.d.). Today Novouralsk covers a territory of 120 square kilometres and has a population of just under 100,000 civilians who have been isolated from the rest of the world for more than fifty years. It remains a city which can only be entered by means of a special pass, or by being listed on the official city roster of ZATO residents, and thus, again, possessing a special pass.

In Novouralsk one can identify three different expressions of the ideal city. The Stalin and early Khrushchev districts built in the 1950s and 1960s; the Brezhnev district of the 1970s and 1980s; and the post-Soviet district of 1990–2000. Each district was conceptualized by the socio-political contexts of their specific times, and each act autonomously separated from each other by tracts of hills and forest.

The first, Stalinist phase, was structured around three main squares, designed in the Empire style of Stalinist architecture, with each square connected sequentially along the axis of a long central promenade that served as gathering places and the route for city parades, creating a sense of unity amongst the citizens through both the ceremony and the formal unity of its architecture. Though highly mannered, this ensemble has a romantic and festive atmosphere while also expressing the emerging wealth of Soviet society and the privilege of the intellectual classes served by it.

The second district was built between 1970 and 1980 to the east, in the era of so-called 'developed' socialism. Following the previous stage of rebuilding the country from the ruins of the war, this was a period when the Soviet Union achieved a relative degree of social stability and a rather high standard of living. This part of the city was conceived as a 'socialist utopia' and highlighted enhanced ideals of social security, wealth and healthy well-being. It was characterized by generous green public spaces connecting retail, health and social facilities, as well as playgrounds and sports centres. All the social facilities were structured along an urban promenade, which bridged over a serpentine road running downhill forming a new local civic centre.

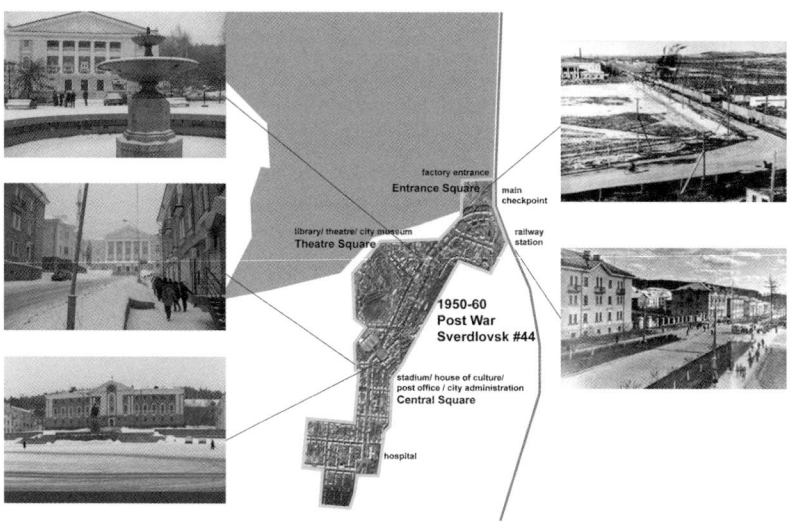

FIGURE 7.1 Sverdlovsk No. 44, the ideal city of the Stalin–early Khrushchev period, 1950–1960s

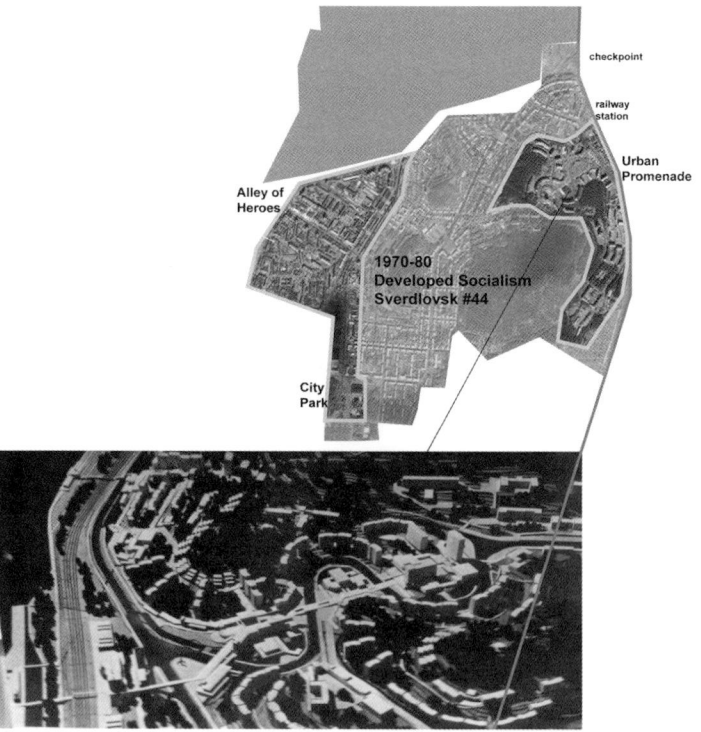

FIGURE 7.2 Sverdlovsk No. 44, the ideal city of the Brezhnev period, 1970–1980s. Project of the central promenade developed by Leningrad Architects, Institute Lengiprogor.

FIGURE 7.3 Sverdlovsk No. 44.

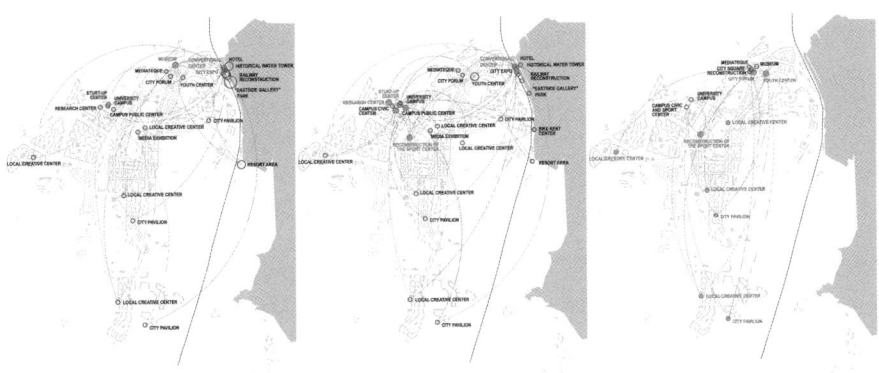

FIGURE 7.4 Workshop Sverdlovsk No. 44, 2011. Working closely with local residents and city authorities our interdisciplinary team together with architectural students developed a dynamic strategy of interdependent initiatives, flexible to re-adjusts according to current socio-economic context and implementation sequence.

This mega-structure of the urban promenade was to be finished with a viewing bridge which penetrated the ZATO wall and made a vital connection between the city and the natural beauty of the lake which lies beyond the secured perimeter, but it was never realized. Another form of the ideal city of the Soviet and post-Soviet period which can be found in Novouralsk is the 'microrayon' – a mono-functional residential district normally containing 8,000–10,000 residents, structured around a hierarchy of primary social and educational facilities. The public centres of the microrayons were initially the schools, usually placed away from the busy avenues and situated in the very core of the district. Today the centre of public life is based upon commercial activity and the public centres of microrayons became 'multifunctional commercial centres'. These 'urban islands' of ideal cities 'have an identity in keeping with their history, social structure and environmental characteristics' (Ungers, 2013, 108). The urban islands within the ZATOs are themselves complete and enclosed enclaves. They are separated from each other by strips of forest, hillsides and remote natural parks, and are connected only by motorways.

There is a very low migration of population within and to the city and the age of the occupants in each enclave is mostly concomitant with the age of each district. Thus the oldest part of the city, officially considered the historic and administrative city centre, is occupied mostly by pensioners, many highly respected citizens who have lived there since the city's construction. This homogeneous demographic is visibly expressed in the city centre, where there is a distinct lack of young inhabitants, and where there is an overall sense of desertion and emptiness. How to breathe life back into this area of the city and to redevelop it presented significant difficulty for my work there.

The fragmented and segregated character of the city instinctively continues in the current city masterplan where the new district is also isolated and connected to other enclaves only by a new motorway, emphasizing ZATOs as a collection of distinguished enclaves, similar to what Ungers was describing in Berlin as 'cities within a city' (2013, 94). The 'urban islands' of ideal cities have given the city a strong identity as a clean, safe, green oasis with a high standard of living. Unfortunately, today the ZATO cannot afford to maintain this lofty utopian image. Having been established, controlled and constantly maintained by the Soviet state, the ZATOs now find themselves under pressure to compete and become self-funding in the context of a free market economy, even though they are still entirely dependent upon constant (though shrinking) central government subsidies. The annual budget of the typical ZATO still exceeds by three times the budget of an ordinary Russian city of the same size. In order to decrease the dependency of ZATOs on central control and funding and make them operate in a more self-sufficient way, the state is gradually transferring the management and maintenance of various social and cultural facilities to the local governments. The challenge for the local officials is great. For the most part, these cities are not capable of maintaining the amazing wealth of social infrastructure given to them – cultural centres, public spaces, sport, health and care facilities. The existing assets are too expansive, and the resources allocated to them are too limited. This difficulty is reflected in the very strange contrast of prosperity and abandonment in the city.

For many years the images of ideal cities from different generations of construction within the ZATO walls contested each other, each one offering an alternative version of the vision of prosperous living. Following this tradition, present-day city authorities are also looking for a completely new image of the ideal place to live, without understanding that the new strategy for ZATO development should not become another template of the 'ideal' city, but rather become a mechanism for the qualitative, incremental transformation of the city. Such a transformation has the potential to gradually erase the notion of the isolation from the social and economic environment of the city, whereas the old, totalizing approach would simply repeat the errors of the past.

The infrastructure of the ZATO

The other ZATO considered here, Zheleznogorsk (formerly Krasnoyarsk-26), which is similar in size to Novouralsk, was also a part of the Soviet nuclear weapons research project. Zheleznogorsk means 'Iron Mountain', and beyond the metaphoric and symbolic associations of strength and loftiness, it reflects the spatial reality of the production and research facilities for the nuclear and space industries that were located here. The city authorities recently approved a new city seal with the image of an atom torn by a bear, which constructs yet another layer of metaphoric and symbolic imagery – here emphasizing that the technological research centre is surrounded and hidden by the wild landscape of Siberian tundra. As with Novouralsk, Zheleznogorsk was built mostly using labour from the gulags.

In Zheleznogorsk, gulag prisoners constructed an extensive infrastructure of nuclear facilities and tunnels carved into a granite mountain as deep as 300 metres underground. This sophisticated secret infrastructure was designed to withstand a nuclear attack and it took fifteen years for 80,000 prisoners to build. Some industrial halls of the plant exceed fifty metres in height. The scale of the system of industrial and transport tunnels is comparable to that of the Moscow metro system, and the comparison is apt, as the tunnels at Zheleznogorsk seamlessly connect to the city with thirty kilometres of railroad, both over and underground (Yamaletdinov, 2007).

Every morning employees of the mining and chemical industrial plant arrive from the districts to the main central railway station, 'Sociogorod' (Socio-city, one of the code names this city had from 1950s to early 1960s), and catch the train which takes them into their workplace deep within the granite mountain. In the evening, at the same hour, they return to the city. Today there are around 5,000–7,000 people employed in the mining and chemical industrial plant contained in the mountain.

The square at the front of the railway station remains empty except for the commuters arriving and leaving at each end of the work day. This part of the city was built in the 1960s and has a spacious boulevard connecting the railway station square to the main city square which contains the obligatory monument of Lenin. The main theatre, in the neoclassical style of the Stalin period, is a free-standing building positioned at the centre of the square. The theatre occupies the place which, in other cities, would be occupied by the building for the local administration. For most of its history the city did not require local administration because all the control was implemented by the federal state (Yamaletdinov, 2007).

Despite the fact that it is a regularly used central part of the city, it does not show any sign of the informal activity typical of such spaces. Similarly, Zheleznogorsk's beautiful and well-maintained squares, boulevards, pocket parks and city parks don't suggest the possibility of flexibility or freedom of use of these spaces and their designs are reflective of predefined patterns of behaviour. For the residents the formal scenography of these spaces, especially the parks and waterfronts, exemplify the city's healthy environment and highlight the background of Siberia's wild nature – and this is part of their collective pride in living and working in an exclusive place. The strong sense of collectivity amongst the residents cannot be underestimated, and it is directly related to the history of being exceptionally well supplied and cared for by the infrastructure of social services during the communist era, and the feeling continues even twenty years after its end. As a vivid example, they still prefer the huge factory canteens to the smaller restaurants and cafes that arose after the period of increasing openness began. Inhabitants of ZATOs are used to the services provided by the state and state industry and they are strongly dependent on the external support. This kind of behaviour, along with the difficulties private businesses experience within the ZATO from the legislative restrictions and segregation of the districts, are among the reasons that local and informal initiatives are missing from the passive public environment of the city. Local residents continue to adhere to the programmes determined for each public space and to the distinct, but already outdated, protocols of behaviour in a secret city. The certitude and stability provided by the removal of individual and collective agency within state-sanctioned structures of society are a comfort to the residents. It thus becomes important to them to retain these structures even as the centralized support for them slips away. The lack of informal initiatives and the passive behaviour of the local residents resulting from this are key reasons why the ZATO public landscape doesn't act to facilitate democratic interaction or engagement. It is not just the formal and ceremonial structure of the spaces themselves, but the habits and frame of mind of the inhabitants that serve to reinforce patterns of behaviour inherited from a legacy of central control.

What must be understood is that the socioeconomic programmes proposed for the redevelopment of the ZATOs must take into account the physical construction of the public realm. Conceived as simplified, legible, machinic spaces built to accommodate a systematized, organized society, the landscapes of the closed city fail to connect actors as agents in public space. In order to provide self-sufficient spaces capable of acting independently, some core conceptions of the planners and local authorities must change. At present the urban landscape is viewed as a passive service or amenity, without its own agency to instigate change. Further, the drive to remake is total, a continuation of the imprinting of utopian aspirations realized as distinct architectural styles from each generation, on cleared ground—the tabula rasa—rather than to remake the city as a work in progress and to reinvest older, ageing phases of development with new life, modes of living, and new and more mixed generations of inhabitants.

Closed and open boundaries

A closed city is necessarily defined by its perimeter. 'Perimeter' is the official term for the ZATO's security infrastructure – the city border. A three-to six-metre high fortified concrete wall that includes within its boundary a military base and several checkpoints. A ZATO's perimeter usually surrounds the city and the secret industry of the ZATO – which itself has its own even more tightly secured perimeter – but it also casts wide to encompass adjacent villages, forests, lakes and agricultural land. Until recently the expenses and responsibilities related to maintenance of the secure perimeter were functions of the state, but now these have been delegated to the city, which has agreed to maintain the infrastructure from its own budget.

The highly visible infrastructure of the secured perimeter makes it abundantly clear that the ZATO's hard edge isolates it and restricts movement and access. It also serves to raise awareness that the rest of the city is composed of such highly programmed infrastructure. The perimeter has a profound effect on social and psychological characteristics of the citizenry it contains, and also has an effect on the form and mechanisms of the city's economy. Several generations have lived their life behind the perimeter wall and as a result a specific ZATO-resident psychology has developed. A majority of ZATO residents are afraid to lose the wall.

During my recent work with one of the ZATOs, our project team were asked in conversations with local authorities and residents to avoid speaking of 'opening up' the closed city. For them it is a symbol not just of protection from crime and strangers (or the illusion of protection, though crime levels are genuinely low in the ZATOs), of their exclusiveness, and of the preferential treatment and comforts offered within. In many ways the wall functions as would a wall around any gated community elsewhere, though complicated, of course, by the official and secret history. The cities' exclusiveness is also not merely that of an economic class or particular lifestyle, but is also bound up with the professional and intellectual nature of the enclave. Within the walls, ZATO residents are 'voluntary prisoners' (Koolhaas et al., 1972).

Residents can be prisoners in other ways too. Well-educated specialists leaving the elite ZATO can find new work elsewhere difficult or impossible to find. While the reasons for this are not immediately clear, it probably pertains to their specialization in industries formerly concentrated in the closed cities, and which have not yet proliferated elsewhere in Russia, and also perhaps that, as a result of long privilege, stability and job security, these individuals lack the initiative to adapt to new work situations or to adversity in general.

The nature of the secure job market within the ZATOs is changing too. In the atomic cities, for example, the secret industry of research, testing and production is responsible for 90 per cent of the cities' productive output (CNCP, n.d.). At present these cities are completely dependent upon this state-controlled industry. Once the city becomes 'open', it is expected there will be a proliferation of other developments such as 'technology parks', which would take advantage of the existing industry and labour pool, generating new ideas and innovative products, thereby bringing an alternative source of support to the city and adding to the overall diversity. However, liberation from the ZATO restrictions is not going to be enough in itself to spark development and innovation. It will be necessary to create a fertile milieu for creativity before new innovations and developments can take hold. In the post-Soviet period, there has been a serious 'brain drain' across the former Soviet states and many highly specialized workers have left the ZATOs.

Another loss, as also experienced in other Russian small cities, is a significant reduction in the youth population and that of younger adults. Though the closed cities provided excellent primary schools, health facilities and a worker-focused infrastructure, creating a comfortable environment for families and the elderly, their environments lack the cultural, educational and social resources to attract motivated young professionals. Facilities and institutions that provide opportunities for interaction, recreation, self-improvement and new initiatives are all vital for this. To survive the process of slow liberation from total state control of the city economy, the ZATO needs to create a new environment both social and physical and an 'active environment where indeterminate processes happen and drives the evolution of the system', or an environment where 'actors absorb, participate and adapt to change' (Sennett, 2006). The city should embrace uncertainty, open-endedness and informality. These are qualities, supposedly intrinsic to the market economy, but also essential to the uncomfortable freedoms necessary to democracy that are currently excluded by the protocols of the secret city.

Richard Sennett in his essay 'The Open City' blames an overdetermined urban environment for many of the limitations and problems of contemporary planned cities. He calls this the 'closed system' or the 'brittle city' (2006, 3). Sennett suggests adapting the concept of the 'open city' into the 'visual and planning tool[s]' used by urban designers (2014). Open systems exist in a dynamic and constant exchange, such as the processes and forces and chance interactions that characterize evolution. 'Chance events, mutating forms, elements which cannot be homogenised or are not interchangeable – all these disparate phenomena of the mathematical and/or natural world can none-the-less form a pattern, and that assemblage is what we mean by an open system' (2006, 6). This dialogic interaction, in which forms, activities, processes and forces are interdependent and relational is directly analogous to contemporary theories of the development of cultural landscapes. The closed system, on the other hand, may seem efficient and stable because it is simplified and legible, but it results in 'frozen cities', which resist growth or change 'since fixed form-function relations make them so difficult to adapt' (2006, 3). The top-down closed city is characterized by zoning, control and management, whereas the bottom-up open city thrives on participation, adaptation and politics. In a 'rather dissonant way', and not to mention disorderly, 'growth in an open city is a matter of evolution rather than erasure'.

For the purposes of adapting the ZATO for a more open future, it must be possible to conceive of how its historic form may be changed incrementally rather than by abandonment or erasure. The fact that living systems may be both closed and open gives hope that the opening up of the ZATOs is feasible. And there are plenty of examples in which both 'planned and organic' urban conditions often exist side by side or are interpenetrating. Addressing this condition, Sennett refers to Stephen Jay Gould's distinction between boundary and border. The first one acts as an 'edge where things end' (2006, 8), drawing impervious walls between organisms and environment, as with the concrete perimeter surrounding a ZATO. Borders imply the opposite – they are places of exchange, like the shoreline of a lake which is a particularly rich ecological niche. Borders act like a cell membrane: they filter and select, they allow access, but they are also partly impervious. The communication that takes place between organism and environment is protective but open.

Novouralsk provides an interesting example of how this openness is beginning to evolve organically along the hard boundary. Despite restrictions within the perimeter zone (already there is a clue that the perimeter is not an edge but a zone), there have been strides toward bridging between closed and open territories. Informally, the curtilage of land adjacent to the perimeter outside Novouralsk has become a place for networking and for pursuing external interests other than those pertaining only to the ZATO. The railway station and the spontaneous market near the checkpoint have recently become public spaces – meeting points, trading spaces, informal gathering places – in much the same way that the same such sites of exchange sprang up outside medieval castle walls. As if to highlight the productive nature of the border condition, an official marketplace in the city centre has been left completely abandoned, while the jerry-built alternative thrives. The informal market outside the perimeter is squeezed into a narrow strip between the railway line, the edge of the lake and the ZATO's wall. Also in the centre of this strip a small hotel is located adjacent to the railway station. The procedure of obtaining a pass to enter the city is lengthy, so this hotel is frequently used for meetings outside the perimeter. The checkpoint through which to enter the city here and the border edge are in a very close proximity to one of the largest districts of the city, and it is also a point of connection between the city and its lake just outside, an important scenic and recreational draw.

This strip of land has very unusual qualities for a closed city. The compression of uses into a limited space creates the kind of congestion that has always been favourable to commerce, bringing people, goods and ideas into close physical proximity with one another and also slowing them down to force engagement. This very actively used space is also a cosmopolitan one. It is a space of communication and trade between two types of actors: dwellers of the ZATO and visitors from outside the wall. In this way, this space forms a porous thick border between both closed and open systems and spaces. The fact that this informal strip of land emerged along the wall proves that the ZATO can act as both closed and open systems capable of exchange and engagement with the outside world through 'bottom up' initiatives. Further, it is possible that, were the city to be thrown open to the outside world in one stroke, this democratic space of engagement might be lost. It suggests that an incremental approach to change would value this space and its past, and use its qualities to project what the city's future ought to be.

Manuel De Landa describes a duality and coexistence of the unplanned and planned and he avers that there is not a simple binary of 'self-organised meshworks of diverse elements, versus the hierarchies of uniform elements' (DeLanda, 2000, 32). He claims 'meshworks and hierarchies not only coexist and intermingle, they constantly give rise to one another. For instance, as markets grow in size they tend to form commercial hierarchies' (2000, 32). This slim strip of the ground along the ZATO wall, allowed a measure of its own agency, has much potential to evolve from informal space of interaction between inside and outside to a space that holds the actual and symbolic power to transform the ZATO, over time, into an open city. To do so, what is needed is adequate strategic support within the social-economic plan of the ZATO development, thereby creating, as Steve Johnson (2002) calls it, the 'controlled emergence' of a new independent city economy combining the potentials of the existing research industry and the associated intellectual capital, and diverse external interests. This notion, however, or any mechanisms for effecting its emergence, is missing in the urban redevelopment strategies of ZATOs, as are the sophisticated definitions of open and closed cities so astutely elucidated by Sennett. Asif Siddiqi, in an article about secret cities, writes that the ZATOs 'were formed originally as sites of knowledge production, but they became, in practice, sites where knowledge disappeared, like absent dots on the map' (2012, 7). The lack of understanding of the crucial differences between open and closed systems also results in the misapplication of models of urban redevelopment. Often strategic urban proposals for the ZATOs are based on case studies of successful engineering and innovation clusters or successful public spaces in ordinary cities, ignoring the fact that behaviour and use of the public realm inside the closed city is very different.

In the project to create a socioeconomic and spatial redevelopment strategy for Novouralsk, I paid close attention to the border condition at the city's perimeter. This active edge became not merely a fortuitous peripheral space, but rather the starting point for a variety of projects proposed by our group. If any of our proposals catches the eye of the local authorities, then they may potentially result in an inversion of the structure of the city's existing landscape, in which the most active business and commercial centre grows from the site of border exchange, not from the existing closed and overdetermined city centre. If the qualities and characteristics of democratic openness exhibited by that border zone can be incorporated into the city's overall structure, then it may become part of its psyche as well, embodying a new spirit of engagement and agency. At Zheleznogorsk we also identified sites where the border condition just before the entry checkpoint might become a new social centre. The gates to the 'shadow city', once symbolic of absolute closure, might now become symbolic of communication and exchange.

The ZATOs are cities with a fragmented character, loose amalgamations of isolated districts composed of images of ideal cities. They are isolated, passive local communities without the means and motivation to communicate either externally or even internally. Taking the ZATO as both a closed territorial formation and a closed city 'mentality' creates a double obstacle to creating agency for the city to interact and prosper in the new social and economic context of today's Russia. Compounding this, we may say that today no one effectively has control over the closed city – the present moment is an interregnum between state and the local control of the citizenry. The state cannot generate the knowledge or interest to attract scientists and young motivated actors to the ZATOs by itself any more, and local residents have never been empowered and motivated to be independent and active actors in the city. They have not been accorded agency in these cities, nor do they particularly welcome it. The informal activity self-organizing along the perimeter of the ZATOs represents the missing environment and the absent agency urgently needed within the closed cities as they begin to overcome their isolation. The border represents the open, the vulnerable, but at the same time the flexible and adaptable, and productive environment. Today the prospect of opening the ZATO perimeter remains fragile, but if the lessons learned at the city's border aren't heeded, then the most likely future will be watchful protectionism, isolation, decay and obsolescence over engagement, agency and economic advancement.

References

CNCP (n.d.) 'United Kingdom–Russia Closed Nuclear Cities Partnership', *Overview* [online], available at: www.cncp.ru/eng/cities/novouralsk/ (accessed 10 April 2012).

De Landa, M. (2000) *A Thousand Years of Nonlinear History*. New York: Swerve Editions.

Johnson, S. (2002) *Emergence*. London: Penguin.

Koolhaas, Rem and Zenghelis, Elia with Madelon Vriesendorp and Zoe Zenghelis (1972) 'Exodus, or The Voluntary Prisoners of Architecture: The Strip', project.

Sennett, R. (2006) 'The Open City'. Available at: www.richardsennett.com/site/senn/UploadedResources/The%20Open%20City.pdf.

Sennett, R. (2014) 'Open City Planning', online interview. Available from: www.youtube.com/watch?v=-LHaKErv7yI.

Sennett, R. (2016) 'The City as an Open System', online lecture. Available from: www.youtube.com/watch?v=ix_QrJUb_Ak.

Siddiqi, Asif A. (2012) 'The Secret Cities', in X. Vytuleva (ed.) *ZATO Soviet Secret Cities during the Cold War*. New York: Harriman Institute, Columbia University.

Ungers, O.M. (2013) 'The City in the City', in Florian Hetweck and Sebastien Marot (eds) *The City in the City, Berlin: A Green Archipelago*. Zurich: Lars Muller Publishers.

Vytuleva, X. (2012) 'The Strong Presence of Invisible', in X. Vytuleva (ed.) *ZATO Soviet Secret Cities during the Cold War*. New York: Harriman Institute, Columbia University.

Yamaletdinov, S. (2007) 'Archivestnik – Krasnoyarsk 26 – Realised Utopia of Soviet Urban Design', online. Available at: http://archvestnik.ru/ru/magazine/ab-6-99-2007/krasnoyarsk-26-%E2%80%93-realizovannaya-utopiya-sovetskogo-gradostroitelstva (accessed 19 November 2016).

8
RHYTHM, AGENCY, SCORING AND THE CITY

Paul Cureton

Lawrence Halprin's (1916–2009) work is part of the canon of Western landscape architectural design and has been widely embraced by the discipline. While Halprin is now part of the roll of famous practitioners, during his working life Halprin constantly looked outside the field of landscape architecture for answers to describe urban environments and ways to understand people's perceptions of environments. Halprin's work, however, has largely been historicized, particularly in the representational tools he developed, and their social agency is overlooked. A series of articles and recent publications have brought his overall work more to the fore, particularly through the scholarship of Alison Hirsch (2014) and Ann Komara (2012). Lawrence Halprin's scores arguably lend themselves to contemporary challenges in landscape representation, particularly in the graphic depiction of time in landscape and in a period where landscape architects are seeking modes of participation.[1]

    Scores for Halprin are based on a musical analogy of a composition which is performed. The elements of the score are location, space, time, people and activities, amongst other things (Halprin et al., 1999, 44). This chapter discusses three discrete modes of work by Halprin, the 'Motation' System, RSVP cycles and Portland Sequence. Motation is a method; RSVP cycles Halprin's holistic approach to creativity and work, and the Portland Sequence is an applied built example of his approach. These discrete parts form a series of works embedded in the popular consciousness of landscape courses and professionals. The aim of this chapter is to connect and realign the valuable agency that this mode of work has achieved, particularly at a time with a resurgent interest in activities which describe the rhythm of places and seek modes to engage people in the built environment. Thus, Lawrence Halprin's work should be part of a process of research seeking to discuss the history and agency of visual representation as a mode of understanding future landscape challenges (Cureton 2016). Agency, after Alfred Gell (1998) and applied by James Corner (1999), should be understood through the life of its representation, from inception to its realized form, impact and use. This undertaking cannot be exercised here in full; a thorough archaeology is necessary, but we can discuss some of the ideas and drawings which Halprin produced as a basis to re-map the dynamics of his realized designs and through such a process understand the mechanics of agency. A thorough and dynamic account of agency and the small stages described which form part of a larger vector, provide a definitive understanding of the roots of landscape design individually and socially, as well as a reflective device of our relationship and understanding of nature. There are not many sources that provide this sort of data return, and an account of the responses to places, the rationale behind design concepts as well as their mode of development as too much of landscape is temporal and fleeting.

Such a realignment of Halprin's work and understanding of agency emerges from the processes of his drawing. Influenced by Paul Klee's *The Thinking Eye* (1961), amongst other sources, Halprin's sketches such as his RSVP diagram mediated between individual responsiveness and reflection, and also as a device for participation. These sketches transpose to design work and can be seen in the Portland Sequence, of walkways and plazas, eight interconnecting blocks of naturalized space and fountains (Source Fountain, Lovejoy Fountain, Pettygrove Park and Ira Keller Forecourt Fountain). These spaces are designed and 'choreographed' for movement which also features quiet and soft nodes for contemplation (Figure 8.4). Jim Burns in his essay on Halprin's drawings (Halprin, 1972) describes how drawings allow a metaphysical leap, in which Halprin exorcised his anxieties, experiences and stresses, everyday life and response to nature. The use of drawing as an emotive device can also be evidenced in Halprin's reflective autobiography *A Life Spent Changing Places* (Halprin, 2011, 244–245). This reflection resembles an observation by John Berger in which he describes drawing as a mode of 'burrowing in the dark, a burrowing under the apparent' (Berger, 2007, 77). The internal rumination aspect of sketching is one part of the agency of drawing. Drawing also has a communicative mode, for example in the work of Randolph Hester, the landscape architect and sociologist, where drawing in its many forms is part of a process of 'representational acts of representation' (Hester, 2007, 97). Hester uses drawing in workshop settings for an ecological democracy (Hester, 2010) of 'representative representation' which is termed governance of the environment by people, emphasizing direct, hands-on involvement. Halprin viewed drawing as a medium which assisted in his approach to creativity and participation through a scoring system. Halprin's work is part of a wider milieu and period of collaboration, experimentation and response to urban challenges seeking new forms of design and making, and new forms of community and engagement in the creation of places. Halprin aimed to reach consensus and commonality in his projects and drawing was one of the mechanisms to help achieve these beliefs. The aim was simple: to find new ways of describing how designed spaces impact people's movement, with a rhythm and body-centred approach, and this as a by-product showed us how conventional visual representation methods in landscape architecture were too static to describe fluid places.

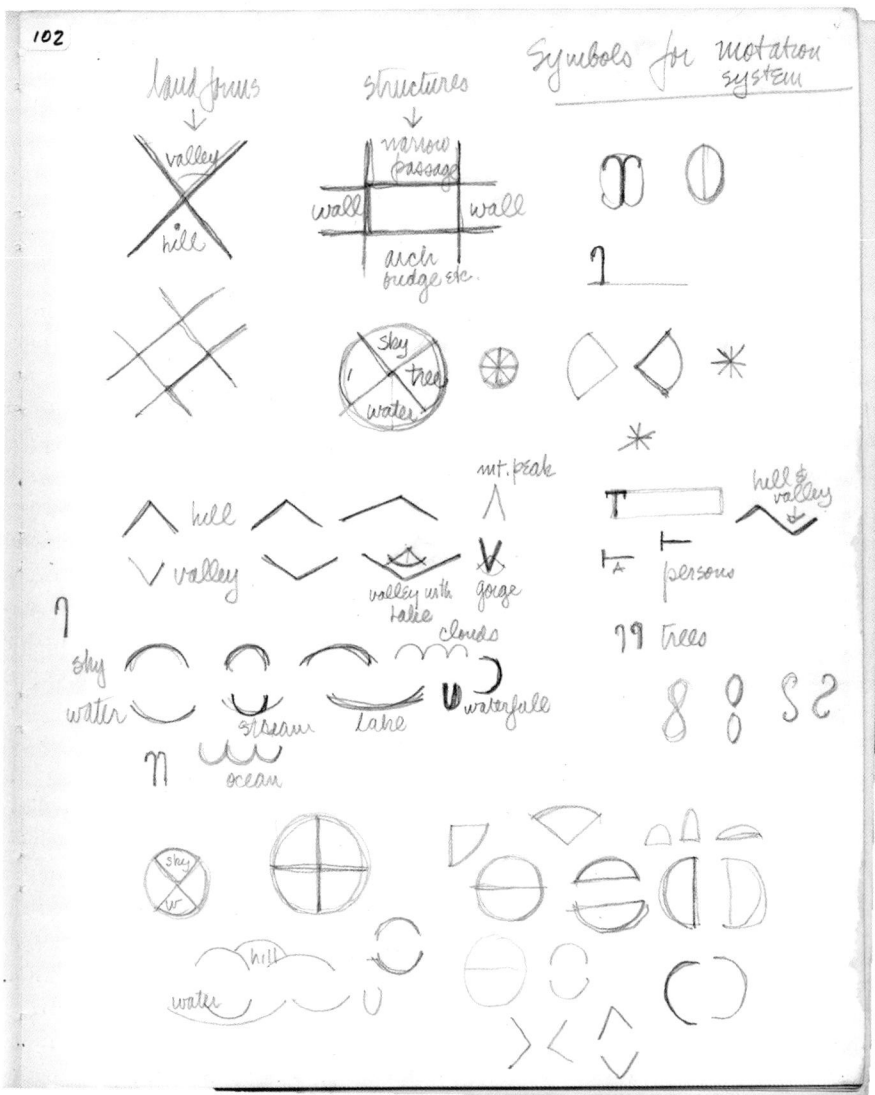

FIGURE 8.1 Notes on a Notation System, University California, Berkeley, graduate seminar, November–December 1964. Image courtesy of the Lawrence Halprin Collection, The Architectural Archives, University of Pennsylvania. See also Motation in *Progressive Architecture* (1965, 126–133).

Motation

> The score is the mechanism which allows us all to be involved, to make our presence felt.
>
> *(Halprin, 1970, 4)*

Lawrence Halprin extended the idea of scores and scoring to include a variety of human activities. By doing so, Halprin, influenced by his wife the choreographer Anna Halprin, created a new representational system[2] that lent itself to community planning consultation and the representation and rhythm of urban places. This system was called 'Motation' – movement and action developed first with dancers and landscape and architecture students. This mode of scoring allowed Halprin to work not just with processes of landscape and of landscape ecology, but also processes of human perceptions and interactions with the environment. The score was often graphic, derived from music and Laban Notation (Laban Movement Analysis or 'written dance', 1928),[3] it preceded the performance, but often the graphic score would become an agent of the performance, it required decoding and analysing. The score could be interchangeable, shuffled and reacted to by diverse groups; it choreographed people in spaces and choreographed spaces for people. In the University of California, Berkeley seminar notes (November–December 1964), Halprin develops the graphic score and notation system, as seen in Figure 8.1, on the philosophical basis that there has not been a system to describe human motion in landscape and urban settings. The graphic device borrows the musical score sheet and acts as a track frame to record the symbols of movement. The frame track indicates distance and speed travelled largely through the process of walking. The notes continue with reference to Laban notation, and also owe a citation to Christopher Alexander (1977). Halprin develops a series of quickly delineated symbols that describe physical static objects, natural phenomena, transport and people; in other words, actions and perceptions of the environment. Scores consisted of a track sheet laid vertically or horizontally. The sheet was divided into a key frame, or overall map of the space being notated, a horizontal track and a vertical track. The vertical track consisted of additional divisions with a distance unit and time unit indicating rhythms of speed and progression. Thus, the two tracks provide a 3D environment to which the scorer can map their body movement and perception. Motation records movement and sets it in a durational sequence. An additional symbol system accompanied the track which symbolized various static, fixed and moving elements of the space. The Motation system could be changed prior to performance or direct performance (Halprin and Burns, 1974, 243). Consultation could add new symbols and elements particular to the location, or the score could be a closed score concentrating on a particular element such as fountains (Halprin, 1963, 134–161) or set out a particular itinerary and series of actions for participants to undertake (Halprin, 1970, 157). Motation requires some directive to the activity, and it was Halprin's environmental philosophy that was the underlying driver. As a device in its own right it is limited due to requiring a necessary attachment to other scores, prefacing, gathering additional information and setting activities towards design plans, however, Motation arguably still has value, as it provides perceptual experiences of places that produce a sort of soft data.

The paper notation as an object has little aesthetic value in itself, but the reflective quality that emerges from participants when each Motation score is compared and commonalities are found is difficult to achieve from other means of representation. These common elements may include the experience of variations in speed due to unfavourable walkways, pollution or traffic, urban signage, the density of places and places of rest. Gathering such reflectivity and response from a site makes these soft data invaluable. It is to be noted that commonality can be construed as too close to Halprin's search for gestalt universals, though when analysing the Motation score reversing the process to find a difference in environmental perception has a unique role. Through drawing symbols on a musical track sheet, the process provides a valuable feedback loop on how the participant's body is moving in space creating novel paths as well as being directed by designed space.[4]

The score could also be mapped to masterplans, elevations and other modes of graphic representation such as sequential sketches similar to the Townscape approach of Gordon Cullen (1961).[5] Motation places movement at the core of the design process, as Halprin states,

> Since movement and the complex interrelations which it generates are an essential part of the life of a city, urban design should have the choice of starting from movement as the core – the essential element of the plan. Only after programming the movement and graphically expressing it, should the environment – an envelope within which movement takes place – be designed. The environment exists for the purpose of movement.
>
> *(Halprin, 1972, 208–209)*

Halprin continually developed the scoring process applying it to a variety of briefs, the Motation method particularly translating into the design of the Skyline Park. The score may also be prefaced by a master score which co-ordinated the overall consultation and design project. As a representational tool, Motation requires a period of learning and acclimatization to the process, most evident in the Motational work of Halprin's contemporaries Phillip Thiel and Kevin Lynch. This process is even more difficult to deliver in contemporary workshop conditions in which a preference for digital mediation overrides in the author's experience of delivery and guidance. Motation's positive attribute is its agency in heightening the participant's environmental perception, though its reflective quality has to be carefully curated and choreographed as the participant's aesthetic and urban prejudices requires further workshops and scores to guide the process forward. Scores as part of the RSVP cycle are the areas in which the fiercest and most justified criticisms have been discussed questioning the potential for manipulation and its true participatory value (Hirsch, 2014, 262–263).

RSVP

Scores in themselves required a more complex system, as Halprin reflected, to describe an approach to creativity. Scores were opened up to explore communities and to the wider city and its urban form, and his developing notions of ecology formed from the Sierra Club and his relationship with Ian Mcharg. For example, sketches conducted for Skyline Park transposed the arroyos of Red Rock mountain park, Denver, Colorado into the material specification and colour palette (Komara, 2012, 48). This mode of work repeated in Portland, Oregon amongst other places and was embedded within his overall RSVP cycle.

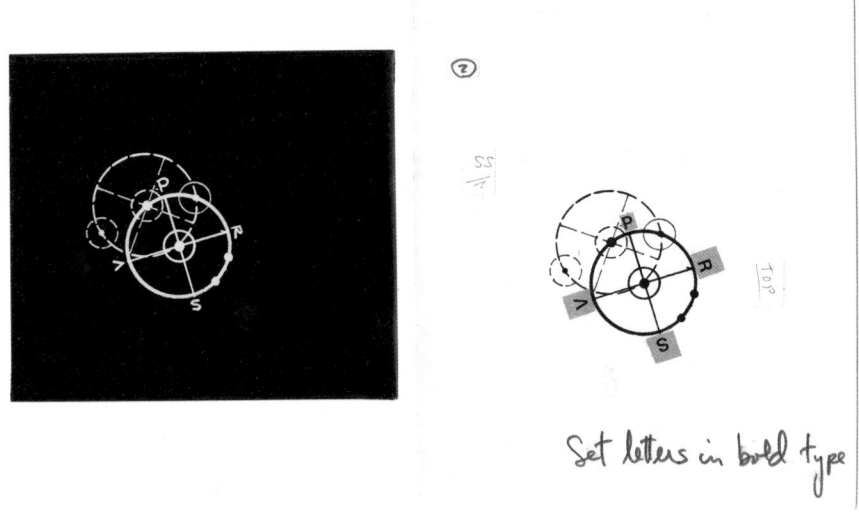

FIGURE 8.2 Lawrence Halprin, RSVP Cycles, 1969. Courtesy of the Lawrence Halprin Collection, The Architectural Archives, University of Pennsylvania. RSVP was an acronym for Resources, Scores, Valu-action and Performance. This multidirectional and overlapping cycle meant that resources were gathered such as human and physical resources, place, economics, emotions, agendas as well as motivations; scores which created the frame for the performance; valu-action is the analysis of and feedback of actions and decoding of findings; and performance is the result of the scores and style of activity to be undertaken. The original composition of the diagram appears on the RSVP book jacket.

Elements of the RSVP cycle could involve Motion scores, interviews, participant reporting scores, plan scores and an overall master score which layer the Resources, Scores, Valu-action, and Performance cycle. The various sites and applications can be seen in the RSVP Cycles (1969), Taking Part: A Workshop Approach to Collective Creativity (1974) and Halprin's book *Cities* (1963 and 1972). Thus, Halprin's RSVP cycles were a mode of recognizing the limits of graphic scores, but situating graphic scores in a wider context so that its agency could be understood in his approach to creativity. The RSVP cycle diagram could be started at any point and has two levels, an inner circle of self-orientation and an outer community orientated circle (Halprin, 2011, 132–133). The diagram repeats, emphasizing the cyclical nature of Halprin's approach to design, signifying a continual process or connection for further projects. As Kathleen John-Alder reflects, the RSVP cycle was the instigation of system theory and feedback loops for guiding, conveying and interacting with scores in the environment (John-Alder, 2012, 65). In this choreography, Halprin's criticisms fall to the use of feedback loops and scores as proactive tools and elements of control towards a preconceived outcome.

As Hirsh states, this outcome was predominately driven to re-conceive a participant's environmental values and Halprin's belief in human universals and commonality (Hirsh, 2014, 263–264). Resolution workshops and interjection scores would be deployed as devices to reach consensus. In this sense, the drawings produced throughout the process were not valued aesthetic pieces executed by an artist's hands. The drawings are embedded with other data forms and participants' work, given order, overlapped, reflected upon and fed forward to new scores, and this is one of the most interesting examples of the communicative agency of drawing in landscape architecture. Thus, drawings produced in the RSVP cycles were de-aestheticized, through a visual application of the Ockham's Razor principle, a sort of economy of drawing[6] and writing codified to provide a common language to participants. This drawing/writing symbol system is reminiscent of earlier sophisticated writing systems such as those found in the tribes of Nsibidi, southern Nigeria (Macgregor, 1909), primitivized by colonial writing.

Portland

The Portland Sequence, Oregon[7] owed its creation to Halprin's early participation with the Sierra Club, Nevada, in group camping trips in which Halprin developed his understanding of nature (Halprin, 2011, 100–101). Halprin's sketches were by and large notated, often with reflective remarks on processes and experiences, while sketching was executed in pen and wash, or watercolour. In some cases, these were photocopied and hand-coloured for exhibitions. The notation is embedded in perspectival work, for example, the various words for water movement such as 'leap, bounce, bubble', etc. (see Figure 8.5). The sketchbook notes his twofold response to Portland, creating a block sequence for movement, and second, to abstract natural phenomena, and then to transpose this to the site. The sketch would not be the only container for ideas; clay models and other techniques would hone the proposals. In his sketchbooks, Halprin draws various cross-sections to demonstrate the role of water in the shaping of the landscape, how the river creates its basin and sediments, how the geology forms underwater currents. From these observations, the idea of erosion and rock form are used here as a basis to draw Halprin's ideas for a fountain as a theatre space. The washed ink perspective sketch was the agent that mediated between the experience of mountainous landscape and its ecological and geological form and the commission to create new localized contexts in designed spaces. This mode of design development was not participatory but situated within the RSVP cycle as an indigenous resource.

In the Portland sequence, the work features water bowls and a plane area for contemplation signifying the catchment and change of flow of the Columbia River and Cascades in the auditorium forecourt (Ira Keller Fountain). Halprin intended to choreograph the soundscape of the forecourt from his observations executed in pen of the medium of water. It can be evidenced in the sketch for Tuolumne River (Figure 8.4) where Halprin notes the elevations and ledges of the pool. Concrete planes are drawn, creating a stepping sequence to the basin. The rocks are derived from his abstraction studies and re-assembled as a choreography of

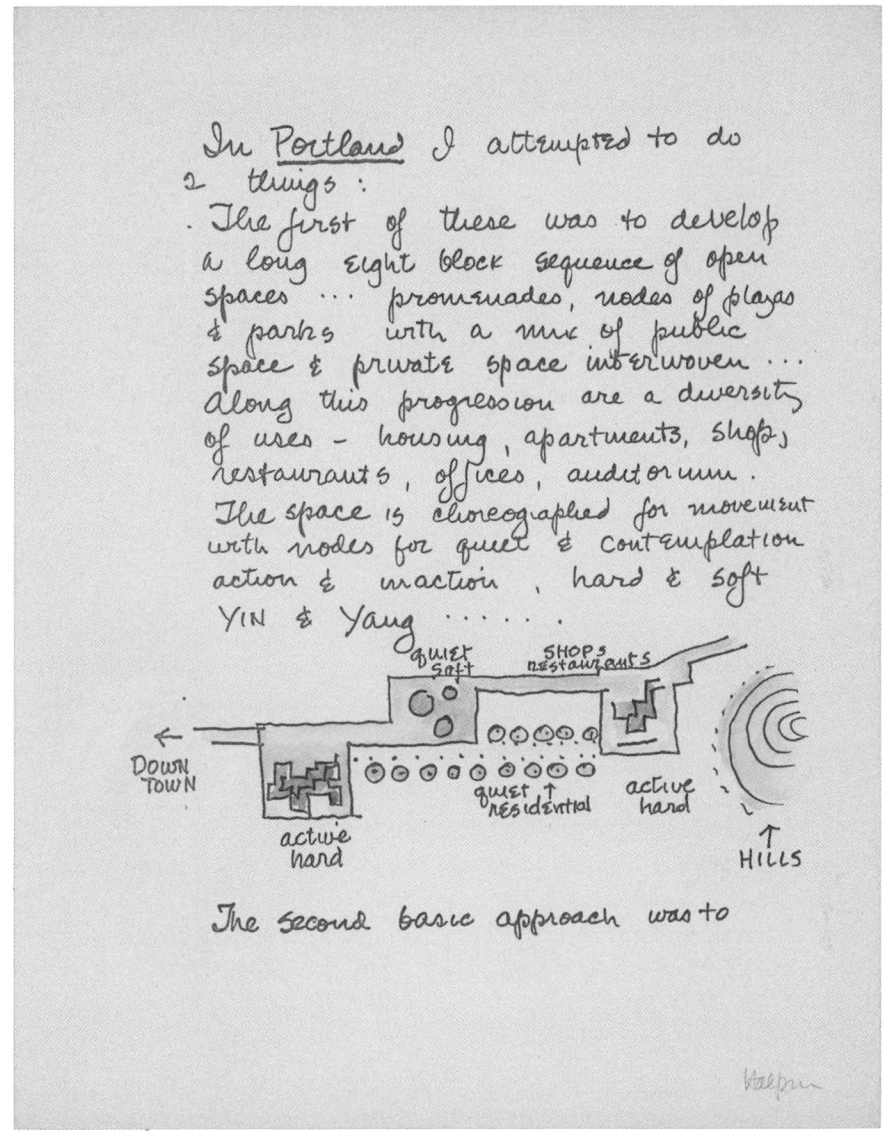

FIGURE 8.3 Portland Open Space Sequence Notes, 1967. Published in Sketchbooks of Lawrence Halprin (1981), p. 61. Image courtesy of the Lawrence Halprin Collection, The Architectural Archives, University of Pennsylvania.

water movement. Halprin's built work, as Randy Gragg comments, brought 'play' back to the city of Portland through its urban renewal program (Gragg et al., 2009). The agency of the drawings can be seen in the embrace of the sequence and their contemporary interpretation in the 2014 Portland Design Week, the support of the Lawrence Halprin Conservancy and Cultural Landscape Foundation and its addition to the National Register of Historic Places.[8]

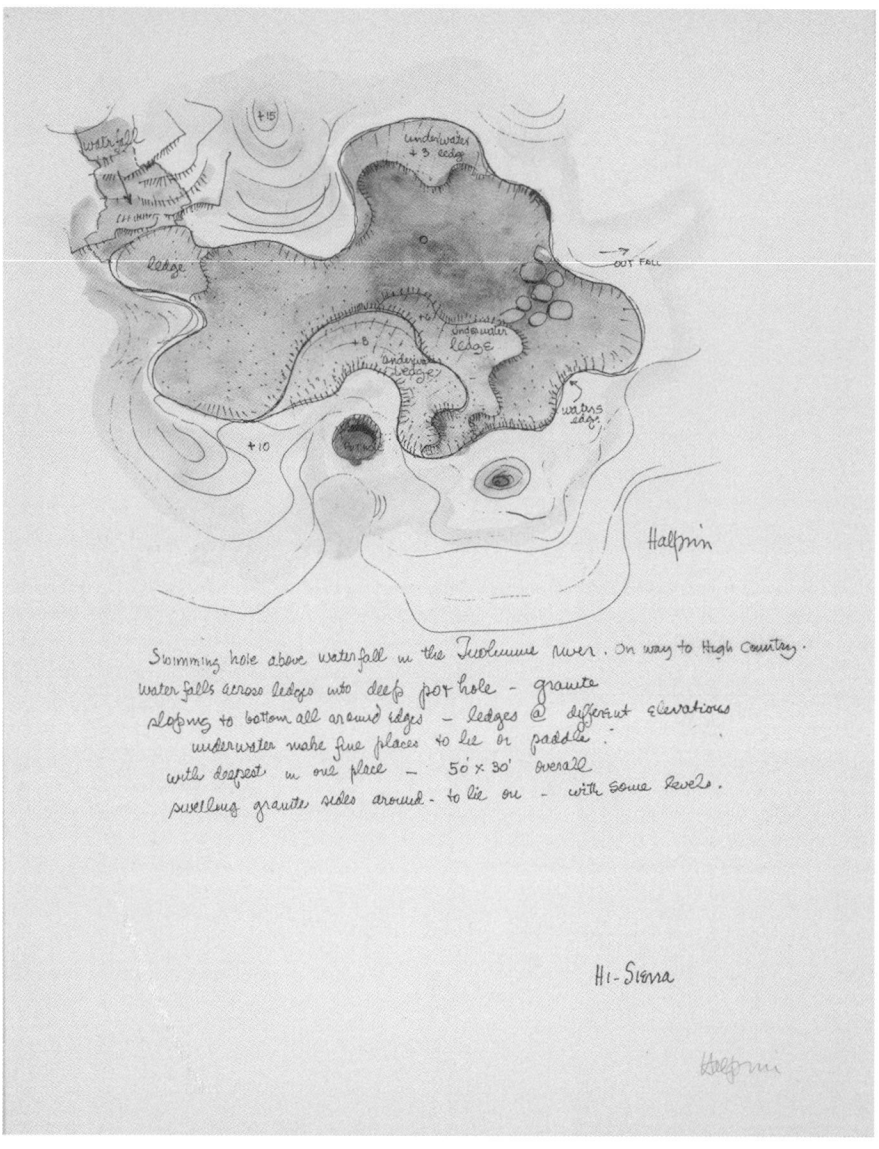

FIGURE 8.4 Portland Open Space Sequence, Tuolumne River, 1962. Published in Sketchbooks of Lawrence Halprin (1981), p. 67. Courtesy of the Lawrence Halprin Collection, The Architectural Archives, University of Pennsylvania.

Summary

The search in our time is for valid processes, and our urban forms will evolve and change as part of our process of development and in response to the changing technological discoveries of the future.

(Halprin, 1963, 9)

FIGURE 8.5 Portland Open Space Sequence, Notated Observation of the High Sierras, 1962. Published in Sketchbooks of Lawrence Halprin (1981), p. 64. Courtesy of the Lawrence Halprin Collection, The Architectural Archives, University of Pennsylvania.

As Randolph Hester reflects on Halprin's work, only theory and practice in concert create useful theory in landscape architecture (Hester, 2012, 142). The diagrams, scores and drawings are evidence of Halprin's attempts to apply his modernist universalisms, a desire to reconnect people with natural spaces and to be engaged with wider planning processes. Motation arguably still carries currency and agency as a representative device in its ability to raise perceptual awareness of environments and increase environmental literacy. Motation has a participatory value in its ability to generate soft data, as it is not gathered by other means; it raises participants' perceptual awareness of the movements of their bodies and the dictation of pattern of movement by design spaces. It also provides a durational aspect, of an action performed in time. The ability to process those data further regarding consultation and design development is still fraught with difficulty. The agency of Motation is reliant upon a will to find new ways of doing things, of collaboration, and this is not easily resolvable, though its process of environmental awareness has a valuable agency. Halprin's view of ecology and the understanding of processes in landscape and their ephemeral qualities transpose from his perspectival notated sketches into the design and specification of his projects (Halprin, 1970, 98). Sketches provide the resources for designed work, unlike Motation. These two modes of operating were part of a wider RSVP cycle and approach to creativity. Such a method of transposing sketches and using experiences of nature as a basis to create metaphorical designs is questionable, though the sketches' value and agency fall into the way that drawing and notating were immersed within participatory workshops, and the drawings themselves were not regulated as singular objects but placed in a wider cycle of production. For our landscape designs and architecture of the future, we need to understand agency, through further critical accounts of the production of images in landscape architecture and intents. In addition to this, modes of creativity that move beyond the domain of images, that describe acoustic, tactile descriptions of places and movement should continue to be understood. Agency projects ideas, though it is also an archaeological device to trace back the social-cultural period in which representations emerge and lays bare who and for what purpose designed work is for.

The social agency that the designed spaces of Halprin and the participatory workshops achieved are significant as part of a period of environmental activism and for the impact on planning processes and urban regeneration in the United States. The contemporary practice of landscape representation, from student to practitioner alike, if it truly re-presents, requires revisiting historical innovations that have been historicized and realigning them for future urban challenges. In other words, landscape futures are reliant on agents that can enact such participatory processes, find commonality and difference and Halprin was one such person committed to this endeavour.

Notes

1 As seen in the works of landscape practices: Scape, Olin, Balmori, Civic Design Lab amongst many others and publications (Murnaghan and Shillington, 2016; Corkey, 2016).
2 Anna Halprin created a new project with the composer Morton Subotnick, 'Parades and Changes' (1965 and 1970) after five years of experimentation in collective scoring using everyday movements and settings (performance by San Francisco Dancer's Workshop at the Stockholm Statstheater, 1965; San Francisco Dancer's Workshop for the opening of the UC Berkeley Art Museum, 6 November 1970). See also Worth and Poynor (2004).
3 Rudolf Laban developed a notation system to record movement of actions, patterns, floor space and bodies in space. The symbol system records different directions in space, with shading indicating the level of movement. The duration and quality of movement are also

recorded. For practical instruction, see the Dance Notation Bureau (DNB) at http://dancenotation.org/.
4 As Halprin states when devising Motation, 'I could compose scores where the notes or symbols denoted both sound and action. Anna could use the system for choreographing her performances and happenings, and I could use it to develop scores for designing spaces and environments' (Halprin, 2011, 128). Motation was a unique device that strengthened collaboration with his wife and provided options for Halprin's firm to take further planning and transportation clients.
5 See The Charlottesville Motation Sequence sketches by Norman Kondy of Lawrence Halprin Associates.
6 This economy of drawing principle also reappears in Catherine Dee's *To Design Landscape: Art, Nature & Utility* (2012).
7 Four connected blocks reflecting the Columbia River and Cascades through the medium of water: the Source Fountain, Lovejoy Fountain, Pettygrove Park and Ira Keller Forecourt Fountain.
8 SEGD Portland Chapter; Mayer/Reed, Portland State University Graphic Design Department, Sticky Co, PNCA Animation Arts, 7 October 2014.

References

Alexander, C. (1978) *A Pattern Language: Towns, Buildings, Construction*. New York: Oxford University Press.
Aronson, S. (2013) 'Lawrence Halprin: Another View', *Landscape Journal* 31, pp. 219–226. doi:10.3368/lj.31.1-2.219.
Berger, J. (2007) *On Drawing*. Aghabullogue, Co. Cork, Ireland: Occasional Press.
Burckhardt, L. (2015) *Why Is Landscape Beautiful? The Science of Strollology*. Basel, Switzerland: Birkhauser.
Corkey, L. (2016) 'Reclaiming and Making Places of Distinction in Landscape Architecture', in R. Freestone and E. Liu (eds) *Place and Placelessness Revisited*. London and New York: Routledge, pp. 61–75.
Corner, J. (1999) *Recovering Landscape: Essays in Contemporary Landscape Theory*. New York: Princeton Architectural Press.
Cullen, G. (1961) *Concise Townscape*, new edn. Oxford and Boston, MA: Routledge.
Cureton, P. (2016) *Strategies for Landscape Representation: Digital and Analogue Techniques*. London: New York: Routledge.
Dee, Catherine (2012) *To Design Landscape: Art, Nature & Utility*. London: Routledge.
Gell, A. (1998) *Art and Agency: An Anthropological Theory*. Oxford: Clarendon Press.
Girot, C. (2012) *Landscape Vision Motion*. Berlin: Jovis Verlag.
Goodman, N. (1976) *Languages of Art: An Approach to a Theory of Symbols*. Indianapolis, IN: Hackett.
Gragg, R. (2014) 'The Portland Open Sequence Podcast Series', Lawrence Halprin Conservatory, available at: http://halprinconservancy.org/ (accessed 16 December 2015).
Gragg, R., Ross, J. and Beardsley, J. (2009) *Where the Revolution Began: Lawrence and Anna Halprin and the Reinvention of Public Space*. Easthampton, MA: Spacemaker Press.
Guest, A.H. (1998) *Choreo-graphics: A Comparison of Dance Notation Systems from the Fifteenth Century to the Present*. Amsterdam: Gordon and Breach.
Halprin, L. (1963) *Cities*. New York: Reinhold.
Halprin, L. (1970) *RSVP Cycles*. New York: George Braziller.
Halprin, L. (1972) *Lawrence Halprin Notebooks 1959–1971*, 1st edn. Cambridge, MA: The MIT Press.
Halprin, L. (1997) *The Franklin Delano Roosevelt Memorial*, 1st edn. San Francisco, CA: Chronicle Books.

Halprin, L. (2011) *A Life Spent Changing Places*. Philadelphia, PA: University of Pennsylvania Press.
Halprin, L. and Burns, J. (1974) *Taking Part: A Workshop Approach to Collective Creativity*. Cambridge, MA: MIT Press.
Halprin, L. and Burns, J. (1986) 'Lawrence Halprin: Changing Places', Exhibition at San Francisco Museum of Modern Art, 3 July–2 August 1986, illustrated edition, San Francisco Museum of Modern Art.
Halprin, L., Hester, R.T. and Mullen, D. (1999) 'Lawrence Halprin Interview', *Journal of Places* 12(2), pp. 42–51.
Helphand, K.I. (2012) 'Halprin in Israel', *Landscape Journal* 31, pp. 199–217. doi:10.3368/lj.31.1-2.199.
Hester, R.T. (1982) *Planning Neighbourhood Space with People*. New York: Van Nostrand Reinhold.
Hester, R.T. (1989) 'Social Values in Open Space Design', *Place Journal* 6(1), pp. 68–77.
Hester, R.T. (2007) 'No Representation without Representation', in M. Treib (ed.) *Representing Landscape Architecture*, new edn. London: Taylor & Francis, pp. 96–111.
Hester, R.T. (2010) *Design for Ecological Democracy*. Cambridge, MA: MIT Press.
Hester, R.T. (2012) 'Scoring Collective Creativity and Legitimizing Participatory Design', *Landscape Journal* 31, pp. 135–143. doi:10.3368/lj.31.1-2.135.
Hirsch, A.B. (2011) 'Scoring the Participatory City: Lawrence (& Anna) Halprin's Take Part Process', *Journal of Architectural Education* 64, pp. 127–140. doi:10.1111/j.1531-314X.2010.01136.x.
Hirsch, A.B. (2012) 'Facilitation and/or Manipulation? Lawrence Halprin and "Taking Part"', *Landscape Journal* 31, pp. 117–134. doi:10.3368/lj.31.1-2.117.
Hirsch, A.B. (2014) *City Choreographer: Lawrence Halprin in Urban Renewal America*. Minneapolis, MN: University of Minnesota Press.
John-Alder, K.L. (2012) 'A Field Guide to Form: Lawrence Halprin's Ecological Engagement with The Sea Ranch', *Landscape Journal* 31, pp. 53–75. doi:10.3368/lj.31.1-2.53.
Jost, D. (2010) 'Lawrence Halprin 1916–2009', *Landscape Architecture Magazine ASLA*, pp. 93–111.
Klee, P. (1961) *The Thinking Eye: The Notebooks of Paul Klee*, ed. Jürg Spiller. Lund and Wittenborn: Humphries.
Komara, A.E. (2012) *Lawrence Halprin's Skyline Park*. New York: Princeton Architectural Press.
Macgregor, J.K. (1909) 'Some Notes on Nsibidi', *The Journal of the Royal Anthropological Institute of Great Britain and Ireland* 39, pp. 209–219.
Murnaghan, A.M. and Shillington, L.J. (2016) *Children, Nature, Cities*. London and New York: Routledge.
Rainey, R.M. (2012) 'The Choreography of Memory: Lawrence Halprin's Franklin Delano Roosevelt Memorial', *Landscape Journal* 31, pp. 161–182. doi:10.3368/lj.31.1-2.161.
Robertson, Lain M. (2012) 'Replanting Freeway Park: Preserving a Masterpiece', *Landscape Journal* 31, pp. 77–99. doi:10.3368/lj.31.1-2.77.
Walker, P. (2012) 'Lawrence Halprin & Associates, 1954: A Brief Memoir', *Landscape Journal* 31, pp. 29–32. doi:10.3368/lj.31.1-2.29.
Wasserman, J. (2012) 'A World in Motion: The Creative Synergy of Lawrence and Anna Halprin', *Landscape Journal* 31, pp. 33–52. doi:10.3368/lj.31.1-2.33.
Worth, L. and Poynor, H. (2004) *Anna Halprin*. London and New York: Routledge.

9

PUBLICITY AND PROPRIETY

Democracy and manners in Britain's public landscape

Tim Waterman

Oxford Circus

There was a most peculiar form of catharsis when the streetscape at Oxford Circus was renovated. Lengthy barriers that would trap pedestrians either on the sidewalk or in the street were removed, and a central reservation to allow refuge from cars was added. People could now dart across the street when an opportunity arose. I use the Tube station there every day, and the change was liberating, like removing a constricting or chafing garment. Indeed, it was a lesson in how the artificial measures used to control pedestrian behaviour can disempower people, can cause an unease that is an often unacknowledged but constant irritant – pedestrian underpasses are a prime example, or the 'pig pens' that confine walkers in highway medians on 'pedestrian refuge islands'.[1] Despite the fact that Oxford Circus is a defining space of consumerist distraction, it was noticeable immediately after the refurbishment that people were more aware, not just of the space, but of themselves and others. The intensity of use in the space is a great leveller. People from all walks of life walking here find themselves both equivalent and responsible for one another. It is a democratic space.

In this chapter I will make the case for the importance of treating the public landscape as a space of democratic engagement, and, by extension, of its importance in the moral life of civil society. I will argue that manners are the expression of virtues and that as such they are an integral part of our shared morality. Design and planning can support or deny ethical encounters in public space, and I shall seek to identify and evaluate some ways in which this occurs to provide some useful examples for design that encourages democratic life in a healthy civil society.

Manners and civil society

The misunderstanding or misapplication of public manners – the set of customs that eases people's negotiation of public spaces – creates problems for the design of public landscapes such as streets and squares. In Britain in particular, much pressure has been brought to bear on design for the public landscape[2] to provide visual cues for behaviour, usually with very mixed results and a preponderance of signage. Despite the British reputation for politeness, the relative absence of customs for appropriate *public* behaviour creates problems not just for design but for the comfort and safety of individuals, and also for civility, recognition and democracy in society as a whole. I don't wish to imply that public manners have necessarily been eroded over time – in many places the public landscape is a more civil place than ever before in history – however it is important to address what *should or could be* in order to provide most fully for democracy and human flourishing.

The first value of the public landscape is that it should be equally accessible to all, regardless of any individual's membership in any minority group and regardless of any person's class. Equality does not automatically confer justice, but in this case the link is fairly clear, as I shall discuss below. This particular equality is primarily guaranteed by manners rather than legal enforcement. Manners, the expression of virtues, are the first and most basic expression of morals as applied in everyday life – and indeed form the ground from which many moral judgements are constructed. Peter Johnson concurs, noting that 'without civility as a minimal condition of human contact . . . principles of justice and welfare would have little permanence and reliability' (1999, xi), and John F. Kasson in his study of American manners writes that manners 'are inextricably tied to larger political, social, and cultural contexts and . . . their ramifications extend deep into human relations and the individual personality' (1990, 3). The construction of public manners is also highly relational, contingent upon place, time and what actors are engaged in a situation. The word 'situation', in and of itself, is telling, and points to the fact that our moral lives are always situated. Here it is important to stress that this is not a position of moral relativism; in fact, it is quite the opposite, as what is sought is a condition of public interaction that assumes certain universal moral goods, such as the avoidance of harm and/or pain, and the aspiration to human flourishing.

Democratic public life depends upon a customary compact between citizens; an agreement as to what is proper in a public context. The notion of propriety has long been associated with sanctimoniousness, of 'polite society'. When manners are constructed of hierarchical relationships, they inevitably involve deference and condescension as well as inflexible and possibly harmful codes of honour. When manners are seen as the foundation of just relations in civil society, then 'polite society' might be seen as simply the mutual regard necessary to ensure the movements necessary to society, a structure in which we are all inevitably embroiled. Put simply, propriety is a form of ownership;[3] individual ownership of the self and its relation to the public world at large. Propriety is defined by custom, and custom, at its best, is not a dogmatic and inflexible framework, but rather it informs and is formed by everyday life and everyday practices. Further, what is proper to society is the mutual recognition in the public realm that is the first gesture of reciprocity – a sort of gift from one person to the next, however small. The anthropologist Stephen Gudeman is one of our finest commentators on this. He writes, 'Strictly a secondary and composite phenomenon, reciprocity is not the core of society but its expression. Anthropological theories have it backwards: reciprocity is neither a primitive isolate nor the atom of society but its badge' (2001, 92). Reciprocity also takes us outside of our communities, into realms of exchange outside of the exclusive bounds of community. Gudeman, to some extent, indicts the construction of communities: 'A commons is regulated through moral obligations that have the backing of powerful sanctions. But communities are hardly homes of equality and altruism, and they provide ample space for the assertion of power and exploitation from patriarchy to feudal servitude' (2001, 27). Much political rhetoric has stressed the importance of community, but civil society is a broader realm than mere community and requires broader consideration. What happens on a sidewalk reflects our understandings of public engagement beyond the bounds of community.

The River Can

For several years my commute, which began at Oxford Circus, took me out of London to the Essex village of Writtle, where I taught at the Writtle School of Design. I would ride the train to Chelmsford from London and then mount my bicycle to complete the journey to work. The last leg of my journey was a very pleasant 3.5 kilometres along a planned 'green finger' along the River Can, a tributary of the Chelmer, for which Chelmsford is named. The path is very attractive, winding through meadows and overhung with mature trees, and it is heavily used even at its iciest or most flooded – indeed these times were sometimes the most beautiful, when the floodwaters would pool across the wide, misty meadows and when the path itself was obscured with a sheet of flowing water. Apart from the aesthetic pleasure of the journey, there was a real sense of security and community afforded by the regular interactions with other path users; waves, smiles and pleasantries. The path was marked at key points with signage indicating that it was a shared surface, but otherwise there were no markings, and the space of the path required negotiations between pedestrians, powered wheelchairs, cyclists and a variety of other wheeled, human-powered conveyances, maintenance vehicles, and, of course, species other than humans – usually dogs. For such a circumstance in Britain there are no set protocols in law or code, and customary use of the pathway had sorted itself out naturally over time and through practice. For the most part, people would keep to the left of the surface and traffic flowed quite easily at disparate speeds. This simple custom, arrived at over time, afforded each person the maximum possible freedom and agency on the path. When the occasional walker would hug the wrong side of the path, it registered as an odd annoyance, but aberrant practices could also present a real danger, when, for example, a runner might come around the inside of a blind corner directly, and at speed, into the path of a cyclist. Over the years I had several frightening encounters and was either the recipient of or was involved in a couple of minor injuries despite the fact that I travel on my bicycle at what a colleague jokingly calls 'parade pace'.

Sidewalk ballet and sidewalk moshing

I have had many discussions with newcomers to Britain who are struck by the difficulty of navigating the sidewalks. An Australian colleague, for example, was accustomed to the general rule that one should keep to the left on the sidewalk (in Australia vehicular traffic keeps left, as it does in Britain), and further that it is rude to force someone to walk nearest to traffic if they have their back to it. In Britain, this rule did not seem to apply, though the latter part of it *does* appear in the Highway Code, the official government guide for all road users (which I will address more fully below). 'Where possible, avoid being next to the kerb with your back to traffic' (Department for Transport, 1995, 5). Unfortunately, this treats the pedestrian as *private* and solely responsible for his or her own conduct. In practice, people walk to the left or right as they will, with many who are able to exercise their dominance preferring the inside of the sidewalk away from traffic at all times, some trying to keep left as a rule, some trying to keep right as a rule, and others ducking and diving to find the path of least resistance. For someone who knows what an impressive lubricant to pedestrian passage such a simple rule can be, its practical absence is a source of constant frustration. It makes one constantly aware of the multiplicity of unnecessary micro-aggressions that comprise life as a British pedestrian. On a relatively uncrowded surface people have ample time to adjust to one another's relative positions, but on a narrow sidewalk in a crowd it is either the most aggressive pedestrian or the person who can project the greatest air of entitlement that wins. Many Britons with whom I have spoken about this profess not to notice; that it is a fact of life and beneath notice. These 'soft forms' of domination (in many ways ethical and aesthetic expressions of denial of respect or esteem, and assaults against self-respect and self-esteem) (Sayer, 2005, 16) are, however, very much indicators of the existence of a pecking order, and are likely to be far more apparent and obstructive to those lower in that order. Different classes (and, indeed, anyone outside the dominant social order for reasons of ethnicity, sexual orientation, and so on) acquire different skills that shape the postures and gestures with which they inhabit the world. Bourdieu calls this the *hexis*, and Tim Ingold insightfully puts it in context:

> Such skill is acquired not through formal instruction, but by routinely carrying out specific tasks involving characteristic postures and gestures, or what Bourdieu calls a particular body *hexis*. 'A way of walking, a tilt of the head, facial expressions, ways of sitting and of using implements' – all of these, and more, comprise what it takes to be an accomplished practitioner, and together they furnish a person with his or her bearings in the world.
>
> *(Ingold, 2000, 162; Bourdieu, 1977, 87)*

The *hexis*, then, is that set of skills and the personal bearing that might allow a person to project an air of entitlement and thus to claim a disproportionate share of public space at the expense of their social 'inferiors'. It is useful to imagine a scenario in which the *hexis* might determine differential rights to space. A person with an upright bearing, head held erect, perhaps wearing a suit and tie, would clearly have an advantage in establishing ownership of his chosen space on the pavement, while another, slouched with hands in pockets, shoulders and head lowered, might be forced to give way. Each actor is expressing the expectations their social position and upbringing has instilled in them, and each is perpetuating a situation of inequality; specifically, that the 'posher' actor has a greater share of personal freedoms that the other. Andrew Sayer examines the place of shame in these kinds of interactions, and notes its importance in determining social order, 'for through it people internalize expectations, norms and ideals, and discipline and punish themselves' (2007, 95). Sayer also points out that 'French working men . . . were less likely to feel shame than their US counterparts because they had a more structural and politicized understanding of class' (2007, 96), which underscores the importance of knowing one's place as one which is just and equal, which I discuss below.

Some with whom I've spoken about this have expressed dismay that I would dare to suggest that personal freedoms could be infringed upon by the 'imposition' of public manners. This points to at least two contemporary problems: one that competitive individualism has become the foundational ethic for much human engagement; and two, that class has become a marginalized subject in Britain, even to the point that official narratives point to the emergence of a 'classless' society. The markers of class may have shifted in British society (for example, being exhibited through 'lifestyle choices'), but the existence of class stratification is as tangible and damaging as it ever was.[4]

'Telling people how to walk is simply not British,' a particularly banal article on public etiquette on the *BBC News* website tells us (Easton, 2014). This is a classic assertion of *doxa*: '[s]ystems of classification which reproduce, in their own specific logic, the objective classes, i.e. the divisions by sex, age, or position in the relations of production, make their specific contribution to the reproduction of the power relations of which they are the product, by securing the misrecognition, and hence the recognition, of the arbitrariness on which they are based' (Bourdieu, 1977, 164). Here 'Britishness' is held up as the unassailable reason that movement and interaction in public landscapes cannot be re-examined. Britishness is presented as intractable and timeless. The mere fact that the author has found it necessary to address the issue serves to highlight that there is a genuine problem. The resultant situation ensures that there is a constant differential between the agencies of citizens in public spaces in Britain that is constantly reinforcing inequalities and injustices that are based in class, lifestyle and other (particularly visible) constructs. Other public behaviour, however, is consistently trotted out exemplifying Britishness, despite Britain's lack of a monopoly on such behaviours. Queuing, for example, is a specific way in which egalitarianism is realized in the British landscape. Queues 'are one way of demonstrating a commitment to the equal standing of other people' (Stohr, 2012, 12). Winston Churchill, a class-conscious Tory if ever there was one, hit out at socialism as having created a culture of queuing, calling socialist Britain 'Queuetopia'[5] (Moran, 2007, 62).

In order to reconcile these frameworks, they must be seen not as problems of Britishness as much as they are purely problems of democracy. We need manners to underpin democracy, and we need the right manners. We need propriety without oppression. 'Established codes of behavior have often served in unacknowledged ways as checks against a fully democratic order and in support of special interests, institutions of privilege, and structures of domination' (Kasson, 1990, 3). To know what is 'proper' is not an antique frippery (or, at least, it shouldn't be), rather it is key to creating democratic civic spaces – spaces where the commons can continue to emerge. Democratic society (and design for public spaces) cannot succeed in the public landscape if people don't *know their place* as a *just and equal place*. If all people are aware of their ownership of their particular place, it will, given time, shift everyday practice, custom, and the *hexis* of individuals.

Goodge Street

My neighbourhood has been gentrifying at breakneck speed. Perhaps most gentrification occurs with such rapidity. When my partner and I moved in, our apartment block was a mix of people, some in rent-controlled flats, like Florrie upstairs, who has lived in the block for fifty years. There were actors working in the West End living four to a room so they could walk home late after the theatres closed rather than taking the night bus to a cheaper, but more remote suburb of London. There were a couple of brothels – one sandwiched between our flat and Florrie's, where Lucy, a pre-op transsexual who had undergone some quite flamboyant plastic surgery in Thailand, turned tricks all day, but never after 10 pm so as not to disturb our evenings. Above Florrie is a prominent neoliberal economist and academic who lives alone. Since we moved in, the price of a flat in our block has nearly doubled, and our neighbours are quickly becoming more wealthy, more bland and more anonymous. The change is noticeable on the street. Fitzrovia, as the area is called, is much busier, especially during the day. The narrow sidewalks along Goodge Street have filled with tables for swanky cafes, and are more crowded with pedestrians. It is now more difficult to walk along the street for this reason, but also for another. The new residents and the visitors to the neighbourhood who are anxious to display how well-heeled they are by mixing here aren't predisposed to give way for others. Groups with linked arms will force people off the footway and into the street (this is known as the 'sidewalk zamboni' after the machine at a skating rink which clears the ice). Couples will walk abreast without yielding any space. Big-shouldered suits yell into their phones and gesticulate while taking their half out of the middle. There's a term for him too – he's called a 'meanderthal'.[6]

Landscape, civil society and the common law

Manners which reinforce class hierarchies, sexual or racial divides, or normative heterosexuality have been consistently challenged over many years, though clearly many negative affectations persist. Here the negative connotations of the term 'know your place' are evident. Tim Cresswell writes, problematically, that to know one's place is more than simply spatial:

> Implied in these terms is a sense of the proper. Something or someone *belongs* in one place and not in another. What one's place is, is clearly related to one's relation to others. In a business it is not the secretary's place to sit at the boss's desk, or the janitor's place to look through the secretary's desk. There is nothing logical about such observations; neither are they necessarily rules or laws. Rather they are *expectations* about behavior that relate a position in a social structure to actions in space. In this sense 'place' combines the spatial with the social – it is 'social space'.
>
> *(Cresswell, 1996, 3)*

The problems here are first that there might be a wish to establish a logical basis for such interactions. Such relations are emotional, moral, practical and often reasonable, but not necessarily rational or logical. Each person has their own desk and their own space that is allocated to them, and it is reasonable to expect a space that may be owned, that is personal. Even in a hypothetically horizontal, non-hierarchical society, each person might still know their place, as a person with particular specialist knowledge or skills, which requires a particular type of space. Teachers and doctors need private offices for consultations with students or patients, for example, while a secretary might be required to inhabit a semi-public space to engage visitors and to serve as a gatekeeper. These roles are profoundly different, and there is no reason to deduce any injustice arising from this particular spatial differentiation. It would be unreasonable for the secretary to occupy the doctor's desk as much as it would be unreasonable for the doctor to look through the secretary's desk. But the doctor's personal requirements are different from the secretary's because their roles are different. In the public landscape people have different personal needs too. A person in a wheelchair, for example, takes up more space on a sidewalk and cannot react as quickly to obstacles as others might. Thus they require a greater level of care on the part of other citizens who encounter them.

What I am proposing is not a return to some form of natural law whereby civil society is ordered according to an atavistic vision of how things 'ought' to be, but rather a vision of society that values publicity as something that is both situated and embodied, collectively and mutually subjective. Civil society, then, is constructed of the interconnections of shared localities and of the sum of them: the ideal civil society, as Michael Walzer says, 'is a *setting of settings*: all are included, none are preferred' (Walzer, 1990, 5). Richard Sennett's work reinforces this, and in his book *Together*, his definition is worth quoting at length, particularly as it stresses the interdependence of civil society, land(scape) and the commons:

> The common law of the land is rooted in custom, which is an expression of community practice . . . It is because custom is rooted in this 'common usage' for 'time out of mind' that custom 'lies' upon the land. The word *law* derives from the Old Norse *liggja*, meaning to lie, and is akin to the plural of *lag*, meaning 'due place, order'. The law, this suggests, was laid down, layer-like, through practice, thereby establishing a sense of emplaced order – the lay(out) of the land. It was in this way that customary rights in the land, such as rights in the commons, created a sense of belonging to, and having a place in, the land.
>
> *(2012, 252)*

Civil society, in order to function well and democratically, requires of at least a majority of citizens that they are capable of exercising publicity, usually within the bounds of propriety, and always in a specific place. Publicity is a personal quality, wherein the individual is able to understand that there is a difference between acting publicly and privately, and to conduct themselves accordingly.[7] Propriety begins with the ability to define the self and its boundaries in relation to others, and not necessarily within a hierarchical social order. Propriety is the understanding and awareness of what is one's own, and in etymological terms it is firmly linked with property, both in terms of possessions and of real estate. So much so, in fact, that in feudal times there would have been a conflation of physical property and personal propriety: 'where political status and authority had to do with family heritage, position in a hierarchy of landholdings, and inalienable connection to a (generally) male-controlled estate' (Davies, 2007, 12). We need now to negotiate what propriety should be in a public landscape in which we may identify as 'commoners' as the first place of individual and mutual empowerment. Again Michael Walzer: 'Civil society is sufficiently democratic when in some, at least, of its parts we are able to recognize ourselves as authoritative and responsible participants' (Walzer, 1991, 303).

In Britain's consumerist society, the balance between perceived personal rights and public agency and the conflict between perceived personal rights and personal agency create numerous problems for the understanding of and design for the landscape of the public realm. It may be that the more complex and challenging that doxic landscape forms take, for example, the proliferation of signage and an uncertainty about public roles and behaviours, the more they limit our ability to challenge those doxic forms themselves, and to see and make alternatives. Obfuscation overwrites the possibility for innovation and democratic interaction. There is also a resistance of the dominant classes to any challenges to a status quo that could be seen to serve them. Pierre Bourdieu notes that the 'dominated classes have an interest in pushing back the limits of *doxa* and exposing the arbitrariness of the taken for granted; the dominant classes have an interest in defending the integrity of *doxa* or, short of this, of establishing in its place the necessarily imperfect substitute, *orthodoxy*' (Bourdieu, 1977, 169).

The space of the street has always been a great leveller, and it takes immense expense and effort to undo this function. The huge investment in the design, construction and ongoing policing and maintenance at the assiduously sanitized and homogenized privately owned public space (POPS) at Canary Wharf is a canonical example of this strenuous undoing. Peter Linebaugh vividly evokes the emerging segregation on the historic London street in his book *Stop Thief!*:

> The street was part of the urban commons. It was not only the place of traffic, or the movement of commodities. It joined producer and consumer, and it joined the producers of various components in separated workshops. It was the site of sport, of theater, of carnival, of song. The cries of London street sellers provided a permanent part of the sounds of the city. By venerable urban custom the puppeteer could set up his Punch-and-Judy show in the middle of the street. The street was erotic, the streetwalker a synonym for the sex worker. Along with the street was the evolution of the sidewalk or 'pavement.' Wheeled and foot traffic were demarcated corresponding to a division between 'economy' and 'society' or between economic production and social reproduction.
>
> *(Linebaugh, 2014, 27)*

Highway codes

Individuals, isolated in either the sanitized POPS or by a doxic landscape that guides their every move by design, become needier and less empowered, and thus better consumers.[8] This forced removal of cooperation and shared know-how has deleterious effects on the negotiated public landscape. Each individual, shorn of obligations to and expectations of the populace around them, must negotiate public space with little, if any mutually constructed guidance. Thus, it is easy to perceive of public life as one in which dog eats dog rather than one in which all individual dogs benefit from the mutual aid provided by the pack. One has the sense that the public would have known what was meant when, in the 1935 edition of the Highway Code, it exhorts road users, 'As the manner in which you use the road affects a large number of others, show care and courtesy at all times' (Ministry of Transport, 2012 [1935], 2). In the 2007 edition this statement is one of the very few to have been simplified rather than expanded, and it says merely 'always show due care and consideration for others', though now without any hint of why, and without reference to courtesy (Department for Transport, 2007, 5).

The Highway Code is one of the few places where agreed manners are disseminated uniformly across British society. It contains rules for manners on roads and paths, but comes into conflict with itself. It treats pedestrians as *private*, but motorists and cyclists as *public*, which results in unique conflicts and significant discomfort in negotiating passage, though few Britons realize the full extent of the problem due to their acclimatization to the existing condition. Though all vehicular modes of transportation are provided with clear rules in regard to their interrelation on the roads, they are merely asked to give priority to pedestrians. Pedestrians are, with only one or two exceptions, expected to behave however they wish, which has the result that their behaviour is utterly unpredictable, inconsistent and selfish. They are private, competitive individuals in a public landscape, in relation to other modes of traffic as well as in relation to one another. The privacy of the pedestrian thus results in a struggle of primacy and deference, or kowtowing and condescension that renders passage a tense ballet. This situation has been further compounded in recent years by an increase in bicycle use, the provision of shared surfaces, and the ubiquitous public use of hand-held and wearable mobile devices. It is also more possible, now, to project an air of disinterested entitlement and opt out of public interaction by focusing intently on one's smartphone rather than one's obligations to other citizens, for example.

I should make it clear, here, that I am not making the case for a set of robust, incontrovertible rules governing pedestrian behaviour, but rather that there should be suggestions for possible behaviour in a given circumstance, and that designers should work to ensure that public landscapes are not overly prescriptive and inflexible. For example, it makes little sense to ask all pedestrians to keep left on a wide pavement on a shopping street with many groups of people walking at different paces and in different directions. This is where rules of thumb are useful – when approaching someone directly, err to moving to the left. An absolute rule would merely become political correctness, where noble intentions are made laughable by rigidity.

The lack of street etiquette creates problems for the design of public landscapes such as streets and squares, and in particular for tricky shared spaces such as towpaths, where tensions between pedestrians, bicyclists and dog walkers can and do escalate into violence. Designers are commonly asked to make public landscapes 'legible', to provide visual and spatial cues to guide users.[9] This can work quite well when built typologies direct people to spaces designed for specific uses, such as those apparent at a park gateway, a railway station or a pub, for example, but there are few ways to help guide pedestrians to behave with regard to one another, short of the very basic elements that indicate the boundaries of a pedestrian realm such as kerbs, railings and planting. It is one thing to employ markers in the landscape to guide legibility and behaviour, and entirely another to shape and modulate the full vessel of public space to the same ends. Legibility depends upon the common use of a shared language, not merely upon the visual cues a landscape can provide, and public democratic etiquette in Britain is anything but a shared language. Designers simply can't program public spaces for gracious human interaction without such language being in place. At present, despite many calls for the reduction of street clutter, the answer is to provide automated signals and posted signage, such as the ubiquitous 'keep left' signs in the London Underground (though sometimes these inconsistently direct commuters to 'keep right'). Millions of gallons of paint poured out to divide lanes could be saved by the proliferation of simple customs for conduct.[10]

A variety of measures that would greatly improve the comfort and aesthetic beauty of our common urban landscapes could be put in place if simple and uniform rules for basic public etiquette were put in place. Further, there are implications for all road users. If, for example, pedestrians were asked to keep left as a general rule, this might make it possible to desegregate all shared cycle and foot ways, including towpaths, and minimize friction between users. Pressure could also be brought to bear to eliminate one-way systems for motorized vehicles as they are often hazardous to pedestrians.[11] If the same simple customs should apply to pedestrians everywhere, a similar simplified set of customs could apply to other modes. Then these conventions could be communicated to the public through such publications as the Department for Transport's 'Stop, Look, and Listen' website for children, tourist guidebooks and the 'Life in the UK' manual which provides guidance for British citizenship, as well as, of course, the Highway Code.

In fact, the issue may not be at all a matter of organizing and marking local spaces, but could be regulatory at a much larger scale. A requirement, for example, for all street and roads, wherever possible, to be returned to two-way traffic, could create the conditions everywhere at local level for customary practices to adjust to this more predictable condition, without the requirement for widespread public information campaigns. No solution, of course, can ever be perfect.[12] Cultural recognition is important to a functioning civil society, but so are local, regional and national planning, policy and design focused upon opening up democratic interactions between people in the public landscape. Thinking and acting at all scales at once, or at least a constant shifting between scales, is a standard of landscape architectural design process and practice that ought to inform governmental processes as they relate to civil society.

Hopefully, it is clear that the argument I have sought to make to establish better street etiquette in public landscapes as a way of building stronger civil society is very different from the standard set of arguments encountered about road use. Commonly evidence-based studies and computer models are used to try to predict behaviour so that it may be accommodated.[13] This, though, allows the foundational questions of what our public realm is for, who it serves, and how it functions democratically and for communities to be set aside. We must ask first how our public landscapes serve our highest common ideals, and then work the rest out from there.

Notes

1 See The Department of Transport's *The Design of Pedestrian Crossings* (1995).
2 I use the term 'public landscape' in this chapter to refer to what professionals in the architectures commonly refer to as 'public realm'. These are the outdoor spaces which have the potential to bring public individuals together as a polity, and where human passage and interaction occurs. 'Public realm' in its fullest and best sense refers to a wide variety of types of spaces in which public interaction might occur. It might be neither outdoors, nor publicly owned and/or managed, such as the ticket hall of a railway station.
3 Legal scholar Margaret Davies describes property in terms of propriety, 'the concept and the manifestations of property in the Western liberal context go far beyond legal doctrine, extending to ideologies of the self, social interactions with others, concepts of law, and social concepts of gender roles and race relations' (2007, 2).
4 Alan Warde has written, 'Some sociologists have made strong claims for the new structural role of consumption practice as a central focus of everyday life, a focus in earlier times provided by occupation. In such a view, lifestyle increasingly becomes a basis of social identity, displacing class as the central organizing principle of social life' (Warde, 1997, 7). This is, however, a displacement different from what Warde had in mind. Here we see a breakdown of public life, in that manners served as a restraint on the upper classes in the form of *noblesse*

oblige. The value of *noblesse oblige* is not reproduced socially as part of a 'lifestyle choice'. Even though it may have been patronizing, *noblesse oblige* at least forced an awareness of the lot of those less fortunate, and without it all that is left is crass entitlement.

5 Queuing, though, is actually not always a clear case of egalitarianism. In the Second World War, Joe Moran tells us, they were often viewed as inequitable, as 'working women, the elderly and mothers with babies ... were less able to queue for long periods' (2007, 62). Again, this simply shows how manners are both contingent and situated.
6 There is not room in this chapter to discuss the display of heterosexual privilege. For this, see Lee Edelman's *No Future: Queer Theory and the Death Drive* (2004) or José Esteban Muñoz's marvellous *Cruising Utopia: The Then and There of Queer Futurity*. Munoz writes about negotiating 'the ever-increasing sidewalk obstacles produced by oversized baby strollers on parade' and that 'the sheer magnitude of the vehicles ... flaunt the incredible mandate of reproduction as world-historical virtue' (2009, 91).
7 Nancy Fraser's discussion of publicity and recognition is useful here, and may be found in her important essay 'Rethinking the Public Sphere' (1990).
8 Another possibility for capital then emerges, that people might come to use new, purchasable technologies to carve out sidewalk space for themselves; to defend personal space without the need for contact or negotiation as exemplified by the crude devices that have been developed to prevent people from reclining their seats in commercial aircraft.
9 'Legibility' can have sinister overtones. Simplifying and making legible can be processes of overweening state power that erase local distinctions, identities and situated knowledges. This is discussed to remarkable effect in anthropologist James C. Scott's epochal book *Seeing Like a State* (1998).
10 For an imaginative alternative sidewalk code, see Michael Sorkin's 'Sidewalks of New York' on Lebbeus Woods' blog (2011): '1. The streets belong to the people! 2. So do the sidewalks!'
11 Design for the public landscape that begins with prioritizing the most vulnerable citizens, pedestrians, could completely change how we design for the most dangerous road users, such as drivers of cars.
12 The self-organization of the Chelmsford path that I discuss above is at its most successful during commuting hours amongst the regular commuters and it tends to break down somewhat at other times in the day. This phenomenon is also visible in busy railway stations, which function well on weekdays when they are full of practised commuters who are aware of their movements and their bodies, but are a nightmare to get through at weekends, when they are equally busy, but when tourists and day-trippers are not as self-aware or practised.
13 This is not to say that algorithms and 'big data' cannot be put to use to effect greater agency for citizens in the public realm, but it is very likely the case that if such systems are not conceived, designed and employed within the relational ground of civil society, manners and the overall psychosocial construction of any public, that they will only serve as measures of control, not liberation.

References

Boltanski, Luc (2012) *Love and Justice as Competences: Three Essays on the Sociology of Action*, trans. Catherine Porter. Cambridge and Malden, MA: Polity Press.
Bourdieu, Pierre (1977 [1972]) *Outline of a Theory of Practice*, trans. Richard Nice. Cambridge: Cambridge University Press.
Cresswell, Tim (1996) *In Place/Out of Place: Geography, Ideology, and Transgression*. Minneapolis, MN and London: University of Minnesota Press.
Cresswell, Tim (2003) 'Landscape and the Obliteration of Practice', in Kay Anderson, Mona Domosh, Steve Pile and Nigel Thrift (eds) *Handbook of Cultural Geography*. London: SAGE, pp. 269–281.

Curtis, Neal (2013) *Idiotism: Capitalism and the Privatisation of Life*. London: Pluto Press.
Davies, Margaret (2007) *Property: Meanings, Histories, Theories*. London and New York: Routledge.
Department for Transport (1995) *The Design of Pedestrian Crossings: Local Transport Note 2/95*. London: The Stationery Office.
Department for Transport (2007) *The Official Highway Code*, revised edn. London: The Stationery Office.
Easton, Mark (2014) 'Advice for Foreigners on How Britons Walk', *BBC News Magazine* website, 18 July, www.bbc.co.uk/news/magazine-28352045 (accessed 31 December 2015).
Edelman, Lee (2004) *No Future: Queer Theory and the Death Drive*. Durham, NC: Duke University Press.
Featherstone, Mark (2008) *Tocqueville's Virus: Utopia and Dystopia in Western Social and Political Thought*. New York and London: Routledge.
Fraser, Nancy (1990) 'Rethinking the Public Sphere: A Contribution to the Critique of Actually Existing Democracy', in *Social Text No. 25/26*. Durham, NC: Duke University Press, pp. 56–80.
Fraser, Nancy (2000) 'Rethinking Recognition', *New Left Review* 3, pp. 107–120.
Gudeman, Stephen (2001) *The Anthropology of Economy: Community, Market, and Culture*. Malden, MA and Oxford: Blackwell.
Ingold, Tim (2000) *The Perception of the Environment*. London and New York: Routledge.
Johnson, Peter (1999) *The Philosophy of Manners: A Study of the 'Little Virtues'*. Bristol: Thoemmes Press.
Kasson, John F. (1990) *Rudeness and Civility: Manners in Nineteenth-Century Urban America*. New York: Hill and Wang.
Kwon, Miwon (2002) *One Place after Another: Site-Specific Art and Locational Identity*. Cambridge, MA and London: MIT Press.
Linebaugh, Peter (2014) *Stop Thief! The Commons, Enclosures, and Resistance*. Oakland, CA: PM Press.
Ministry of Transport (2012 [1935]) *The Highway Code* (facsimile of the 1935 edition). London: HMSO.
Mitchell, Don (2003) *The Right to the City: Social Justice and the Fight for Public Space*. New York and London: The Guilford Press.
Moran, Joe (2007) *Queuing for Beginners: The Story of Daily Life from Breakfast to Bedtime*. London: Profile Books.
Muñoz, José Esteban (2009) *Cruising Utopia: The Then and There of Queer Futurity*. New York and London: New York University Press.
Olwig, Kenneth R. (2002) *Landscape, Nature, and the Body Politic: From Britain's Renaissance to America's New World*. Madison, WI: University of Wisconsin Press.
Olwig, Kenneth R. (2005) 'Representation and Alienation in Political Land-scape', *Cultural Geographies* 12, pp. 19–40.
Olwig, Kenneth R. (2009) 'The Landscape of "Customary" Law versus that of "Natural" Law', in Kenneth R. Olwig and Don Mitchell (eds) *Justice, Power and the Political Landscape*. Abingdon and New York: Routledge, pp. 11–32.
Ross, Kristin (2015) *Communal Luxury: The Political Imaginary of the Paris Commune*. London and New York: Verso Press.
Sayer, Andrew (2005) *The Moral Significance of Class*. Cambridge: Cambridge University Press.
Sayer, Andrew (2007) 'Class, Moral Worth and Recognition', in Terry Lovell (ed.) *(Mis)recognition, Social Inequality and Social Justice: Nancy Fraser and Pierre Bourdieu*. London and New York: Routledge, pp. 88–102.
Sayer, Andrew (2011) *Why Things Matter to People: Social Science, Values and Ethical Life*. Cambridge: Cambridge University Press.

Scott, James C. (1998) *Seeing Like a State: How Certain Schemes to Improve the Human Condition Have Failed*. New Haven, CT and London: Yale University Press.

Sennett, Richard (2012) *Together: The Rituals, Pleasures, and Politics of Cooperation*. London: Penguin.

Sorkin, Michael (2011) 'Sidewalks of New York', *Lebbeus Woods* blog, 25 December, https://lebbeuswoods.wordpress.com/2011/12/25/michael-sorkin-sidewalks-of-new-york/ (accessed 15 April 2016).

Stohr, Karen (2012) *On Manners*. New York and Abingdon: Routledge.

Thompson, Peter (2013) 'Introduction', in Peter Thompson and Slavoj Žižek (eds) *The Privatization of Hope: Ernst Bloch and the Future of Utopia*. Durham, NC and London: Duke University Press.

Walzer, M. (1991) 'The Idea of Civil Society: A Path to Reconstruction', *Dissent* 39, pp. 293–304.

Warde, Alan (1997) *Consumption, Food and Taste: Culinary Antinomies and Commodity Culture*. London: SAGE Publications.

Waterman, Tim (2014) 'Pedestrian Etiquette, Gormless Phone Users, and the Rise of the Meanderthal', *The Conversation UK* website, 8 August, https://theconversation.com/pedestrian-etiquette-gormless-phone-users-and-the-rise-of-the-meanderthal-30282 (accessed 31 December 2015).

10

THE POWER OF THE INCREMENTAL

Agronomic investment in Lisbon's Chelas valley

Jill Desimini

I am a visitor to Lisbon. I am not the typical tourist. Instead I have come to see small agricultural projects that occupy the urban interstices – tiny gardens, tended largely by Cape Verdean immigrants, on forgotten lands nested along highways, rails, valleys and other corridors throughout the city. As a visitor, my understanding of the Portuguese *Estado Novo* (New State) is learned rather than lived. I know that António de Oliveira Salazar ruled Portugal for over thirty years, an authoritarian regime that curtailed political freedoms, but of course I do not have a visceral sense of the impact on the social and material urban constructs. Nearly a half century after his death, the physical and economic presence of the past regime, and its colonies, is everywhere in Lisbon, the capital city. Some of my tour guides have remarked that it was easy to build things during the Salazar era, and the many looming buildings from the period reflect this. Their scale and monumentality are a sort of geologic testament to old fascist power; their aging facades and unkempt entrances reflect current unease and economic hardship.

I am struck by the massive weight of the physical constructions on the landscape. Yet, it is not the heaviness of the actual architecture that haunts but rather the way in which that architecture sits on Lisbon's hilly terrain (Figure 10.1). For example, in the Chelas valley, a deep ravine on Lisbon's east side, the massive modernist residential towers from the 1965 *Plano Urbanização de Chelas* (Chelas Urbanization Plan) sit disconnected on the high ground. The valleys between them are rifts, physically and socially. I perceive them, then, as double-deep chasms.

The buildings appear both immovable and tenuous, heavy weights on land that falls away beneath their foundations. They are cold and stern, of varying architectural quality, while the valleys are warm, cultivated and animated by the evidence of active human care. The physical juxtaposition of these two poles – the planned, technocratic building and the spontaneous, incremental use – sparks an unhealthy dialectic. The immobile encounters the mobile. The historical circumstance abuts

Lisbon. City Scale. 5m Contour Interval.
1:60 000

FIGURE 10.1 Valleys of Lisbon. This map shows the relationship between Lisbon's steep topography and its building morphology. The Chelas valley is highlighted on the east (right) side. Image drawn by Tiffany Dang. Data source: Câmara Municipal de Lisboa.

the contemporary flux. The government plan is set against adaptations made by people who are categorically excluded from the planning process yet responsible for the outcome of the planning. Government agency – and the abuse of this power – abuts local circumstance.

The history of Chelas is a mini-urban cycle: from agronomic colonization, through designed habitations with unintended allotment gardens, back to designed agronomic inhabitation. In the end, the fertile soil – and its use – could not be denied. The landscape, itself, has agency. The *Plano Urbanização de Chelas* barely acknowledges this core value of the valley. But ironically, it is thanks to this neglect that the valley lay open for appropriation. In the past few years, these appropriations have been recognized, by the city and through design, as the basis for new ecological parks that reinforce local practice while bolstering environmental systems at the regional scale. On one hand, the planning process has shifted: the government who, fifty years ago, produced and executed an inadequate mega-plan for its underserved constituency, today looks to members of that disenfranchised community – the allotment farmers who have appropriated the unclaimed city land – for new micro-plans to execute within the gaps left by the previous mega-plan. On the other hand, the farmers become agents of government planning. The plan is both theirs and someone else's, given to them, again, to carry out.

It occurs to the visitor that the local residents are stewards of an appropriated landscape, and that their stewardship is oddly powered by the vacuum of alternative forms of stewardship. In many instances, once this temporary landscape is reclaimed by a city for development, the local stewardship is ignored and obliterated. In other words, local power, or agency, only exists when there is no other course of action. When a government bends to sanction volunteer and spontaneous use, it is remarkable.

But it is also remarkable that in a territory planned and designed precisely to exert control over informal occupations of the land, that subsequent informal occupations are driving the future planning and design of the territory. In the space left behind by the singular, grand planning gesture – the bold and shortsighted urbanism of the 1950s and 1960s – incremental land practices have found agency. The architecture has largely failed; its clumsy feet clobbering the terrain. Instead, cultivation offers a means to build on the ingrained richness, rather than the perceived poverty, of the Chelas valley.

Plano Urbanização de Chelas

The *Plano Urbanização de Chelas* was a direct response to a public housing program, *Habitações de Renda Económica* (Affordable Housing), launched in 1959 by the Portuguese *Estado Novo* for 128,000 new inhabitants in Lisbon (Tulumello, 2015).[1] Designed to the eradicate informal settlements cropping up in the city, the program gave the newly created Gabinete Técnico de Habitação (GTH) the authority to plan and build social housing to meet the identified housing crisis[2] facing the city (GTH, 1965; Nunes, 2013). Coming at a time of increased urbanization in Lisbon, the project was a last attempt of the *Estado Novo* to exert control over the physical territory.

Chelas and the surrounding district of Olivais, areas just south and east of the Lisbon's airport, was a loosely occupied agricultural area in 1959. Technically part of the city, this area was considered rural, and therefore, underutilized. Previously, the difficult topographic conditions had prevented major building, but the GTH, armed with the discoveries of modernism and its new political power, was unfazed, and this eastern front of the city was deemed the perfect place for its new housing experiments. On over 737 hectares, or 10 per cent of the city's total area, the municipality planned three contiguous developments: Olivais Sul, Olivais Norte and Chelas.

At 510 hectares, Chelas is by far the largest, most complicated topographically and, to this day, the least resolved of these three developments. Anticipating growth, the municipality began seizing private land in the eastern part of the city for expansion in the 1930s (Nunes, 2013). This included land adjacent to the Chelas project, but did not include the majority of the land within the area of urbanization plan. Instead, the city's slow expropriation of the nearly 250 private estates within the area of the Chelas plan continued into the 1970s (Heitor, n.d.).

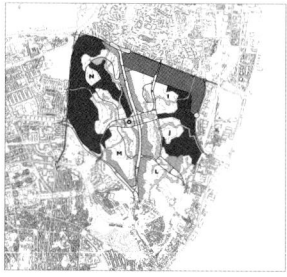

FIGURE 10.2 Evolution of Chelas. This set of drawings shows the Chelas territory before the execution of the *Plano Urbanização de Chelas*, the diagrammatic land use plan of the *Plano Urbanização de Chela*, and the territory after the execution of the plan. Image drawn by Tiffany Dang. Data source: Câmara Municipal de Lisboa.

So slow was this expropriation that the execution of Chelas plan lasted nearly fifty years.[3] During this time, the political, economic and social climate changed dramatically. The *Estado Novo* was ousted. Portugal joined the European Economic Community. Immigration continued to increase, and social traditions shifted.[4] The Chelas project was conceived at the tail end of the functionalist city movement[5] when faith in the modernist project was waning but no subsequent ideals were yet in place, and as such it represents a huge government investment with great social implications executed without a clear or lasting guidance.

The Chelas plan is dominated by the housing-centric, technical concerns of the GTH. It is divided primarily into housing zones (I, J, L, M and N) with a commercial central zone (O) designed to link the housing clusters together into a larger district (Figure 10.2). The plan would seem to resemble lungs, as the activity in the trachea-like central zone seems intended to oxygenate the bronchi-like housing clusters, but the Chelas plan fails to circulate activity through its tissue. Part of this failure is due to the poorly phased implementation of the plan – again executed over decades[6] – part is due to the lack of connectivity and part is due to the lack of communal spaces. Taken together, these oversights represent a paucity of investment in the public realm and a gross misunderstanding of the topographic conditions in the area.

The project provides housing units, and housing units alone. The buildings, aptly, are constructed on the ridges and away from the valley drainage system – but the valleys sit between them with no direct connections across. To move between the zones on the provided routes as a pedestrian or a motorist is nearly impossible. The planners seem not to have fully understood the difficulties that the steep valley sides pose for connection. Instead of adapting the previous grain of development, which provided clues as to how to navigate the topographic conditions, they imposed a cellular system, comprising linear and disconnected developments.

Zona O, the intended nexus, was not realized until 2008. As a self-contained mall with housing towers rather than a neighborhood center, it feels born out of an even more antiquated ideology that an enclosed shopping center could generate economic and urban stability. Again, instead of providing the life the area needed, it only furthered the social and physical isolation. The public transit network is equally meagre, forcing people to walk great distances to have access to the rest of the city of Lisbon. Without these lifelines, the developments are island-like. The unarticulated, dead spaces around the buildings complete the sense of isolation. They are park-less towers-in-a-park, sitting like stale éclairs on a cake plate. The physical development screams of the housing-centric approach of its planners. They say: only housing units are needed to solve a shortage of housing units.

Yet, the developments have plenty of raw open space, which, while city-owned, has no clear assertions of civic ownership and identity. In this absence, the local population has been resourceful in using the spaces for gardening and food production. In the large valley below Zona J, for instance, hundreds of small allotments have emerged in the past decade, interspersed with remnants of centuries-old structures (Figure 10.3). Intrepid gardeners have discovered the dry but farmable soils here, and all along the valleys in the area.

In these spaces, the unplanned movement from housing unit into valley landscape is excitingly alive. It is precisely where the plan left undefined green, that resident-driven ecological and social investment took hold. If you stand with Zona N2 to your back, for example, you see a narrow path deftly negotiating the steep terrain in a way that the four-lane highways that traverse the neighborhood cannot. This is the route from house to garden required for economic subsistence. It is a route that was once taken by squatters, by people who illegally farmed city-owned land. Now it is the route taken to sanctioned *hortas* (allotment gardens) banded together into the municipally funded, designed and operated *Parque Hortícola do Vale do Chelas*.

FIGURE 10.3 Spontaneous *hortas*. This set of drawings shows the rapid evolution of one patch of spontaneous *hortas* below *Zona J*. Image drawn by Tiffany Dang. Data source: Google Earth.

FIGURE 10.4 *Parque Agrícola de Chelas*. These two panoramas, taken from the eastern side of the valley, show the *hortas* before (below) and after (above) the construction of the *Parque Agrícola de Chelas*. Before photo: Câmara Municipal de Lisboa. After photo taken by the author.

From *quinta* to *horta* to Parque Hortícola

The present-day agricultural activities fit within a long history of landscape cultivation in Chelas. Past land colonizers were lured into the valley for its amenable soils just like current inhabitants. They incrementally built up garden estates that capitalized on the favorable growing conditions as well as the topographically enabled views to the Tagus River. These relatively small, enclosed agricultural estates, some with elaborate gardens and houses are known as *quintas*.[7] While particularly abundant in Chelas, *quintas* are a quintessential Lisbon landscape typology (Telles et al., 1997). They have existed in the city for centuries; some have Roman and Visigoth origin, while many were built between the sixteenth and eighteenth centuries. They are part of the rural patrimony of the place, and are fundamental to understanding the way in which the territory was claimed, colonized and cultivated.

In 1959, the Chelas area was covered with *quintas*. Over twenty-five were called out by name on the *Plano Urbanização de Chelas* base map (GTH, 1965). In fact, this was the area with the greatest concentration of extant *quintas* in Lisbon (Telles et al., 1997). The urbanization efforts were not sensitive to this cultural history. The plan did identify key views and buildings to preserve, but the seven clusters were all located to the periphery of the main developments and were not integrated well with the new constructions. Instead, the urban design, starting with the analysis and continuing through the execution, fragmented a continuous terrain, and replaced it with essentially nothing. The spaces between the buildings were left unarticulated, either intentionally through a wise decision to maintain the valleys free of buildings, or unintentionally, through a lack of design of the open spaces within the zones, or both.

Most of the *quintas* were demolished, and only in the *Parque de Bela Viste* was the rural structure deliberately maintained. The other *quinta* remnants are on the valley slopes, and sit, not because they were deemed of cultural value, but rather because their location in areas found to be unsuitable for building somehow spared them. On the sides of the valleys, the organic organization of the terrain – occurring over centuries and at the hands of individuals and small groups of people – presents a stark contrast to the stiff, ahistorical, planned layouts of the housing developments. Here, in the richness of the soil and the deep-seeded cultural practices of cultivation, the legacy of the *quintas* is alive. The actual structures may have disappeared or been degraded, but the human activities persist and evolve.

If the *quinta* represents 'a deeply humanized landscape' (Telles et al., 1997, 161) where different systems of land ownership and exploitation are expressed over a long history, then the *horta* is its postcolonial successor (Figure 10.4). A *horta* is an allotment plot, a small garden farmed for subsistence or pleasure. As previously mentioned, the valleys of Chelas are covered with *hortas*, where residents either squat on unregulated municipal plans to grow food, or, more recently, rent plots, sheds and spigots from the city for personal use.

In the past decade, the city of Lisbon, finally recognizing the cultural importance of urban agriculture, initiated a citywide strategy to organize and facilitate local agronomic endeavors within city-designed, built and managed horticultural parks. The first *Parques Hortícolas* opened in 2011, with the *Parque Hortícola do Vale do Chelas* following two years later. The parks build, conceptually and literally, on the historical *quintas* and the contemporary *hortas*, to provide an armature that promotes the agricultural patrimony of the city.

The *Parque Hortícola do Vale do Chelas* is the city's largest to date with over 200 active plots on nearly six and half hectares. The project, built on the hillside below the *Igreja Santa Clara* (Saint Clare Church) and above the Chelas valley floor, organizes a pre-existing community of *hortas* into an expanded park. The site is an obvious one for the municipal promotion of agricultural activity: it is municipally owned land in one of the important ecological corridors identified by the *Plano Verde de Lisboa* (Lisbon Green Plan) with a rich history of agricultural

FIGURE 10.5 Sheds and teepees. The garden sheds cascade down the valley, in clear lines, while the *cana* bean poles are interspersed throughout the hillside. Photos taken by the author.

use and a pre-existing farming community that depends on the land for food production and economic viability. Further, it sits within the contentious Chelas development where the government failure to provide humane affordable housing to a largely African immigrant population has been well publicized over time. Finally, prior to city involvement, the subsistence farmers had been allegedly tapping into a nearby school's sewer waters as a means of irrigation during periods of drought, giving immediate impetus to restructure the land (Sousa and Batista, 2013).

The project is one of re-organization rather than re-conception. The transformation is subtle but instrumental (Figure 10.5). The re-installation of the *hortas* improves access, public perception and irrigation quality. It is a strategy for land improvement and cooperation, aimed to provide social and physical stability. The moves are simple. The plots have been equalized[8] and arrayed along the hillside in linear terraces descending to the valley floor. Intermittent, evenly spaced, paths cut directly from the top of the hillside to the bottom. For every four gardens, there is a windowless wooden shed; its warm hue glows in the incredible Lisbon light while its red corrugated roof compliments the earthy tones of the dry valley. Adjacent to this shared shed – its use negotiated among the individuals[9] – are rain barrels to collect water from the roof and the water spigot through which clean water flows in times of need, mostly from June through August. The material palette is restrained and controlled: wood and metal sheds; wood and wire fences; small wood retaining elements for the terraces and paths. The same materials are used around the city, and this is a deliberate aesthetic choice, a means to mark the territory as unified and intentionally designed, both on site and across the multiple *Parque Hortícola* projects in the city.

To cultivate the site is to enter into an agreement with the city. For a nominal yearly fee,[10] the water and shed are provided. In turn, the gardeners agree to respect the property and its material palette,[11] maintain their cultivation practice, use water prudently, complete garden training and garden organically. The rules are clear and intended to maintain the communal atmosphere of the *hortas* and to elevate their status within the Lisbon landscape vernacular. The *hortas*, rather than reviled and considered eyesores, are instead celebrated as an organized community embedded within a larger park structure.

The conversion is not perfect. For example, there is now a paved path at the bottom of the valley impeding drainage; and other amenities, including the lighting and the playground, have come slowly and are not fully integrated with the *hortas*. But overall, the *Parque Hortícola do Vale do Chelas* is full of good intentions, supported by physical design interventions and social agreements.

These interventions work on multiple scales: from the individual 160 square meter plot to the clusters of four plots that share a shed; to the larger park that includes skating, play and other forms of recreation; and finally, to the municipal system of green valleys that form the cross grain of the city's ecological plan, the *Plano Verde de Lisboa*.

Plano Verde de Lisboa

The *Plano Verde de Lisboa* (1997), a plan and publication led by landscape architect G. Ribeiro Telles, proposes an ecological structure for Lisbon's metropolitan area. The structure, adopted into the official city master plan in 2007, includes three east–west bands of differing character connected by valley radials. The first band includes the peripheral areas of Monsanto Park, the eastern parks around Chelas, the airport and the remnant agricultural estates. The second band includes the state-owned properties and the public housing developments. The third band is the area along the Tagus River. Radial spokes along the city's main valleys – including Chelas – run transversely through the swathes. In this scheme, the valleys do not just connect the bands. Instead they are fundamental to improving the ecological and social qualities of life in the city. They are conceived of as means to encourage urban ventilation, channeling air currents moving up from the water into the congested areas of development. They are also recognized as the best sites for planting. The protection of the valley soil is an environmental and cultural project. As evidenced by Chelas, the valleys retain traces of the city's rich agricultural history, and still serve as reserves for the current inhabitants.

Since adopting the *Plano Verde de Lisboa* as the aspirational ecological structure for the city, the municipal government has been planning and designing small projects. The approach is incremental, but not unsubstantial, especially given the economic and political climate.[12] In the office of the director of *Direção Municipal de Estrutura Verde, Ambiente e Energia* within the *Gabinete de Projecto de Estrutura Verde* (Municipal Division of Green Structure, Environment and Energy within the Green Structure Project Office) sit over 700 white project binders and a pinkish-red map of the city documenting the work to date on the *Plano Verde*. The map shows the ecological infrastructure projects carefully inlaid into the city. The small fragments aggregate to form interconnected territories within the built fabric. The *Plano Verde* emerged out of a rough tracing of the city, from an aerial image, where the pockets of green were connected and reinforced, idealistically, without direct regard for property, politics and people. Now, the plan is being tested on the ground. The city is a slowly building a series of small landscapes, like inlays, designed to address the environmental – flooding and earthquakes – and socio-economical – food security, immigration and poverty – concerns of the city.

These two agendas – the environmental and social – meet beautifully in the *Parque Horticola* projects. Here, the underlying aim to protect and enhance the soil conditions of the city is infused with a social necessity for subsistence. The planning efforts organize an informal but dwindling reality in Lisbon. The abandoned valley landscapes are being used for cultivation – to produce food for consumption, distribution and sales. At one time Lisbon's topography was maximized for agriculture. Now, it is a practice left to the fertile interstices. In order to augment and valorize current cultivation practices and extend the patrimonial lineage of production landscapes, while providing for safe and healthy growing practices, the city has developed a typological landscape – the *Parque Horticola* – to merge agricultural, recreation and mobility. Skate parks and playgrounds are placed within tracts devoted to gardening. Bike and pedestrian pathways move through the allotment landscapes. And most importantly, fresh water access is provided for crop irrigation.

To a visitor, it is disarming in its clear sightedness. The rural landscape of production, once subsumed by the city, is being given new life. The cycle is direct and evident, especially in the areas surrounding the neighborhood of Chelas, where the agricultural past has not been fully erased, where ancient *quintas* can be seen alongside contemporary *hortas*, where parks like *Bela Viste* have retained traces of their past agricultural organization and parks like the *Parque Horticola Municipais Vale de Chelas* present an agricultural future. In an era of social and economic disjunction, these landscapes provide social, environmental and economic continuity to the city without relying on the category of the nation for direction.

Conclusions: common landscapes and toolkits for mediation

I see the *Parque Hortícola Municipais Vale de Chelas* as a contemporary form of landscape commons,[13] one where land is collectively cultivated through understated means. The cultivation is achieved through the careful organization of quotidian design elements: shed, spigot, fence, path and terrace. These additions, plus a facilitated negotiation process of plot assignment and a simple contract, yield a shared ecological and cultural amenity. The city owns the land but transfers the right of use to local farmers and citizens, including many recent immigrants from Africa (Fonseca et al., 2008). The design of the terrain becomes a form of mediation of both the social and physical landscapes. The allotments are clustered around shared sheds, water and access points. The rows of allotment clusters form terraces to best occupy the steep terrain. Water is managed as it descends into the valley. The public can move through the site, freely, in the spaces between and around the gardens. Occupation is promoted, managed and regulated. Commons have rules – and this contemporary reinterpretation is no exception. It is a right to use the land, but that right comes with responsibilities for both the city and its farmers.

In a near-tragic architectural turn, the project's banality is what makes it remarkable. The design is beautiful without trying to do too much. It is analogous to an introvert whose quiet, listening presence is welcome in a room of loud extroverts jockeying for attention. The design embodies the rough elegance of its terrain. It adapts the intrinsic structure of the previous allotments while respecting the grain of the topographic condition. It generates a symbolic lushness that celebrates the collaboration between city government and city dweller. There was skepticism on different sides of the project – from the pre-existing farmers who feared displacement to the general public who dislike the disorderly *hortas* – but so far, the valley is working as a respected social enterprise. Whenever I walk through the park, I feel a sense of optimism. The farmers are actively tending the plots, the soil manipulated with great expertise. The hillside is lush, as the clean water allows for the crops to flourish in a way they cannot in the plots without municipal spigots. The materials and distribution of the gardens give a sense of unity to the landscape, a legibility that makes me think back to the images of the valley in its pre-urbanization plan state.

This juxtaposition of design and tradition is exemplified in the moments where the cane structures meet the red-roofed sheds. The cane, *Arundo donax*, considered an 'invasive' species in Portugal, is widely used in agricultural practice as a material for building fences and plant support mechanisms. Likely an ancient introduction from India into the Mediterranean basin, the cane has been around a long time, dividing land and supporting legumes. In the Chelas valley, the farmers form teepee-like lattices to support beans, a necessary ingredient for the Cape Verdean specialty *Cachupa*. The beanstalks wind themselves around the cane, overflowing at maturity, signaling bounty, sustenance, life. The cane structures, a similar height and girth as the shed structures, are light, ephemeral and seemingly transitory. By contrast, the sheds represent a more durable stamp on the terrain. On the one hand, they could be critiqued as too fancy for a garden shed. On the other, they signify a proud investment in the land and its farmers, designed to be lasting beacons of a sustained agricultural and social promise. The cane is flexible and accommodating while the wood is a durable anchor. Both constructions signal an incremental investment in human capital that stands in stark contrast to the looming architectural leviathans on the hillside above.

As I stand with my back to the buildings, and peer over the steep edge and into the valley below, I am struck by this marvelously dexterous negotiation of multiple terrains. The bank is so steep that it appears impassable to the non-resident's eye, but the well-worn footpath indicates the contrary. The soil looks dry and infertile but the lush bean and cabbage stalks indicates the contrary. The social environment looks inhospitable but the human activity indicates the contrary. The valley is about the perseverance of the landscape, and of its inhabitants, of its crops, and even of its governing body. All have an inherited agency here. What began as a single-minded endeavor has evolved into a place where – out of neglect – many agents have taken hold.

In facing the valley, it is possible to momentarily forget the past architectural failures of the buildings in the background and see the value of this common project for the city now and in Lisbon's future.

Acknowledgements

I would like to thank Daniel Bauer, Silvia Benedito, João Castro, Tiffany Dang, Maria José Fundevila, João Gomes da Silva, Manuela Raposo Magalhães, Sam Sullivan and Robert Sullivan for their insight and generosity, and the Harvard Graduate School of Design for funding support.

Notes

1 From 1932 to 1950, only 15,904 affordable housing units were constructed in all of Portugal. These units were single-family homes. The 1959 housing act dramatically increased the volume of units constructed, using a high-rise multi-unit housing typology.

2 The GTH conducted a housing conditions survey and with the results, declared a housing crisis in Lisbon due to the high number of illegally occupied and overcrowded units.
3 By contrast, Olivais Norte was planned and constructed from 1955 to 1964 and Olivais Sul from 1960 to 1967 (Toussaint et al., 2013).
4 The democratic revolution of 1974 had unintended consequences for the settlement of Chelas, resulting in a highly segregated community of mostly African immigrants.
5 Here, the functionalist city movement refers to an analytical method, emerging from CIAM 4 'The Functional City' Congress, used by architects and planners to promote the functional separation of urban activity.
6 The housing was built from the 1970s through the 2000s, with differing architectural styles and quality. The earlier developments Zona I (1970s) and Zona N2 (late 1970s) have higher incidence of resident satisfaction than the later developments, Zona N1 (1980s), Zona J (1980s–1990s), Zona M (1990s) and Zona L (2000s) (Tulumello, 2013). Zona J was once considered the most dangerous neighborhood in Lisbon.
7 The word *quinta* comes from the word for one-fifth and refers to the amount of produce, one-fifth of the total crop, paid in rent for the right to farm the land. The word now, is used more generally, to refer to an agricultural estate.
8 Previously, the plot size varied widely whereas the new plots are of equal dimensions.
9 At the onset of the garden plot allocation, pre-existing farmers were involved in the formation of the clusters. Instead of being assigned, the farmers were asked to choose their shed partners.
10 The fee depends on the type of *horta*: for the *horta urbana* where production is either recreational or agricultural, the fee is higher than in the *horta sociais* when use is restricted to meet the alimentary needs of the participant. The Chelas project is made up entirely of *hortas sociais*. The annual rent is roughly €75.00.
11 Materials are limited to wood and cane and gardeners are forbidden from constructing additional structures, including retaining ponds, within their plots.
12 Portugal has been in an economic debt crisis since 2010.
13 The commons is admittedly a complex, loaded and overwrought term, but one that nicely describes something shared and something mundane (Stilgoe, 1982). Its definition is tied to land use rights and property as well as spatial and aesthetic characteristics. It describes land with resource value but should not be considered a synonym for open lands. Commons are constructed and contested spaces, governed by rules with limited shared access. The term is used to describe rights of use on public or privately owned lands, on lands held in common ownership, or on unenclosed lands. The commons can also be a public good or a commonly held commodity but here, the term is being used to refer to the physical landscape.

References

Câmara Municipal de Lisboa (n.d.) *Parque Hortícola do Vale de Chelas: Regras de acesso e utilização das hortas urbanas*. Lisbon: Câmara Municipal de Lisboa.

Câmara Municipal de Lisboa (n.d.) Plan of *Parque Hortícola do Vale de Chelas*. Lisbon: Câmara Municipal de Lisboa.

Colding, J. and Barthel, S. (2013) 'The Potential of "Urban Green Commons" in the Resilience Building of Cities', *Ecological Economics* 86, pp. 156–166.

Fonseca, M.L. (coord.) et al. (2008) 'City Survey Report: Lisbon', GEITONIES, Lisbon, Centre for Geographical Studies, Institute of Geography and Spatial Planning, University of Lisbon.

Gabinete Técnico da Habitação da Câmara Municipal de Lisboa (GTH) (1965) *Plano de urbanização de Chelas*. Lisbon: Câmara Municipal de Lisboa.

Heitor, T. (n.d.) 'Olivais e Chelas: operações urbanísticas de grande escala'. in3.dem.ist.utl. pt/msc_04history/aula_5_c.pdf.

Magalhaes, M.R. (1993) 'Ecological Structure for Lisbon', *Ekistics* 60, pp. 360–361.
Nunes, João Pedro (2013) 'Le Gabinete Técnico de Habitação et la réforme du logement social à Lisbonne (1959–1974)', *Le Mouvement Social* 4(245), pp. 83–96.
Oliveira, V. and Pinho, P. (2010) 'Lisbon', *Cities* 27(5), pp. 405–419.
Pinto, P.R. (2009) 'Housing and Citizenship: Building Social Rights in Twentieth-Century Portugal', *Contemporary European History* 18(2), pp. 199–215.
Sousa, R. and Batista, D. (2013) 'Urban Agriculture: The Allotment Gardens as Structures of Urban Sustainability'. In *Advances in Landscape Architecture*, Edited by Murat Özyavuz, ISBN 978-953-51-1167-2, 938 pages, Publisher: InTech, Chapters published July 01, 2013 under CC BY 3.0 license DOI: 10.5772/51738.
Stilgoe, J.R. (1982) *Common Landscape of America, 1580 to 1845*. New Haven, CT: Yale University Press.
Telles, G.R. et al. (1997) *Plano Verde de Lisboa*. Lisbon: Ed. Colibri.
Toussaint, M., D'Almeida, P.B. and Alcântara, M.D. (2013) *Guia de arquitetura de Lisboa: do movimento moderno à atualidade: 1948–2013*. Lisbon: A+A.
Tulumello, S. (2015) 'Fear and Urban Planning in Ordinary Cities: From Theory to Practice', *Planning Practice and Research* 30(5), pp. 477–496.
Viljoen, A., Bohn, K. and Howe, J.J. (2005) *Continuous Productive Urban Landscapes: Designing Urban Agriculture for Sustainable Cities*. Oxford and Boston, MA: Architectural Press.

Interviews

Castro, J. Interview with author, 10 September 2015.
Fundevila, M.J. Interviews with author, 9 September 2014 and 10 September 2015.
Magalhaes, M.R. Interview with author, 9 September 2015.

11

POST-LANDSCAPE *OR* THE POTENTIAL OF OTHER RELATIONS WITH THE LAND

Ed Wall

Have we reached a post-landscape condition? Have prevailing visual relations between people and land, exemplified by English traditions of pictorial settings, individual perspectives and enclosed properties, reached a conclusion? Has a particular frame of landscape, which Denis Cosgrove describes as a 'way of seeing' (1985, 45), come to a close? Conceptions of landscape, that emerged in fifteenth- and sixteenth-century England and that have continued to be reinforced through contemporary architectural representations and designed transformations, package landscapes as scenic backgrounds and frame tracts of land as spatial products. While referring to these dominant relationships with the land, Barbara Bender reminds us that there are many other ways of conceiving of landscapes: 'when the word "landscape" was coined and used to its most powerful effect, there were, at the same time and the same place, other ways of understanding and relating to the land – other landscapes' (1993, 2). What she describes as contrasting, and often contradictory, constructs of landscape, defined through individual and societal relations with our environments, have grown and receded in relevance. Landscapes are defined through specific economic, social and spatial contexts. So while dominant pictorial ideas of landscape may endure for some people in countries influenced by Anglo-Saxon traditions, other landscapes are configured through contrasting material, ecological, cultural and symbolic relationships with land. In this chapter I explore two inseparable contemporary London landscapes, Paternoster Square and the Occupy London Stock Exchange (LSX). I question a continuation of these English landscape traditions that embrace: predominantly *visual* approaches; scenes considered from static *positions*; and singular perspectives *framed* as representations and urban spaces, enclosed and transformed through design. Raymond Williams proposes:

> It is possible and useful to trace the internal histories of landscape painting, landscape writing, landscape gardening and landscape architecture, but in any final analysis we must relate these histories to the common history of a land and its society. And if we are to understand changes in English attitudes to landscape, in the eighteenth and nineteenth century, this is especially necessary.
>
> *(Williams, 1973, 120)*

Strong associations between the growth of European capitalism in the fifteenth and sixteenth centuries and the emergence during this time of a new way of understanding landscape provide a point of departure for this questioning. Bender states: 'True, the word was originally coined in the emergent capitalist world of western Europe by aesthetes, antiquarians and landed gentry' (1993, 1). The appropriation of what was often common land, and its subsequent enclosure, engrossment and commodification, marked a significant moment in the development of distinctly English landscapes. Kenneth Olwig describes:

> first representations of landscape scenery in painting tended to be views seen from the window of the urban patron whose portrait was being painted and that the same persons who imported Dutch surveyors and engineers to England to restructure and rationalise their properties, imported landscape paintings and hired landscape architects.
>
> *(Olwig, 1993, 332)*

Market relations began to supersede what had been feudal arrangements with the land and its peasantry. Lords became landowners, evicting and relocating populations, collecting rents from tenant farmers and transforming land into property that could be bought and sold. This rupture in social and spatial relations simultaneously facilitated the growth of capitalist economies. Over the subsequent centuries enclosed lands were refashioned as scenic settings for this recently landed gentry. Fences and walls of enclosure were removed or hidden from view (often replaced with ha has) in attempts to control access, dictate views and make claims to extended landscape views of borrowed scenery. John Dixon Hunt's detailed critique of what later became known as the picturesque, explains that the results of 'a growing preference for form at the expense of the ideas that might be expressed through it . . . are unfortunately still with us today' (1994, 16).

A continuation of representing and reconfiguring environments through visual images has explicitly informed contemporary landscapes over the last three decades: these landscape techniques have been employed in the privatization of public spaces; they have contributed to elevated land values; and they have dictated the accessibility, uses and activities of urban redevelopments. As visual techniques from landscape painting are co-opted in architectural computer renderings to facilitate aggressive real estate markets, picturesque approaches continue to misrepresent

FIGURE 11.1 Denied access to Paternoster Square, Occupy LSX encamped at the bottom of the steps of St Paul's cathedral in 2011. Image credit/permissions: Ed Wall, 2011.

how designed spaces will manifest in use and form. These scenic priorities require increased control over urban spaces, which are frequently sought through enclosures and ownership of sites and the subsequent architectural conditioning and management of what activities will be undertaken and by whom. What results is a continued transfer of land to commercial interests and narrowly controlled access to refashioned urban spaces, many which are claimed to be public.

The circumstances around this exploration of landscapes are situated in London, in the aftermath of the 2008 global financial crisis. Bender describes that:

> The way in which people – anywhere, everywhere – understand and engage with their worlds will depend upon the specific time and place and historical conditions.
>
> *(1993, 2)*

Following an economic recession, which saw severe government cuts across the UK, increases in university tuition fees, further privatization of public services and restructuring of state spending, several commentators and critics described the end days of capitalism (Fisher, 2009; also see Harvey, 2015; Mason, 2015). During this time, once accepted crises of capital were questioned as politicians and economists grasped for an understanding of the magnitude of this particular financial turmoil and how to appropriately respond. Although London continued to operate much as it had before the crash, previously accepted economic certainties were severely undermined. David Harvey writes in *Seventeen Contradictions and the End of Capitalism* (2015):

[But] what is so striking about crises is not so much the wholesale reconfiguration of physical landscapes, but dramatic changes in ways of thought and understanding, of institutions and dominant ideologies, of political allegiances and processes, of political subjectivities of technologies and organisational forms, of social relations, of the cultural customs and tastes that inform daily life.

(Harvey, 2015, ix)

Ideologies of capital had been challenged and their relationships with cultural and social practices were strained. Political positions became more polarized and contested, over agendas of economic austerity, policies of accommodating or denying immigration and approaches to public services, such as education, health and transport. Simultaneously, new landscapes of intense control, through security, gating and fencing-off, which had intensified since the privatizations of the 1980s and 1990s, were countered through the resistances of protests and demonstrations. The shutting down of parts of London by landowners and by workers' protests, student sit-ins and collective occupations attempted to redefine social relations through reclaiming public spaces. These contrasting politicized London landscapes undermined scenographic promises of managed, pacified and comfortable urban spaces which had accompanied the marketing of many of the spaces that were reclaimed. New landscapes became defined through mass demonstrations, which gathered in public spaces along Whitehall, in Trafalgar and Parliament Squares, and through the spatial appropriations of sites across the capital. In particular, attempts to demonstrate outside the Stock Exchange in Paternoster Square, defined as Occupy LSX, and their encampment at the bottom of the steps to St Paul's cathedral brought issues of economy, politics and landscape into immediate proximity.

Witnessing claims to the end of capitalism and the emergence of alternative ways of defining landscapes led me to question prevailing landscape practices of urban redevelopment: if capitalism emerged in the fifteenth century (together with scenographic approaches to landscape) – and if this economic model was to come to an end at the beginning of the twenty-first century – would accepted traditions of landscape be correspondingly challenged? Would a post-landscape condition emerge to question a prevalence of visual approaches viewed from ego-centred positions and controlled by static frames? In this chapter I take these spatiotemporal contexts to explore three underlying conceptions of scenographic landscapes: an emphasis on *visual* concerns; static and limited *positions* from which views of spatial landscapes are observed; and narrow, impermeable *frames* constructed to contain and control landscapes. I propose that reconceived landscape relations, which address political, cultural and environmental anxieties, have the potential to correspond more closely with notions of public space. And I advocate that these post-landscapes are not products but are always in the process of being made and remade and where 'people engage with it, re-work it, appropriate it and contest it' (Bender, 1993, 3).

Inseparable landscapes

On 15 October 2011 the Occupy movement gathered outside London's Stock Exchange, in Paternoster Square, to protest against a range of issues including the adequacy or appropriateness of responses to the economic collapse that had unfolded over preceding years. Occupy protests in London reflected corresponding occupations in New York and followed other mass demonstrations in London about student fees and workers' rights. As protestors assembled in Paternoster Square the demonstration was thwarted by police and private security guards. They had been instructed by the square's owners, through a court injunction, to restrict access. Mitsubishi Estate Co., the private landlord of Paternoster Square, continued to exert its control by filling the space with a mesh of steel barricades. Barriers were set up at each entrance with signs proclaiming the private nature of the land and threats of random bag searches and trespass. The public space of the square was closed down as the owners asserted their rights of private property. Signs were erected stating:

> Paternoster Square is private land. Any general licence to the public to enter or cross this land is revoked forthwith. There is no implied or express permission to enter the premises or any part without consent. Any such entry will constitute trespass. Limited consent is hereby given, but can be revoked at any time, for entry on to the accessible parts of the square, solely for access to offices, retail units and leisure premises for genuine building, retail and leisure purposes. Visitors must at all times comply with the directions given by our security personnel.
>
> *(Sign at Paternoster Square, October 2011)*

FIGURE 11.2 Paternoster Square, during the Occupy protests of 2011, was fenced off from undesirable protests, people and activities. Image credit/permissions: Ed Wall, 2011.

As the Occupy protesters were blocked from bringing their concerns to the door of the London Stock Exchange, they found refuge in an interstice between St Paul's cathedral and the exterior building walls of Paternoster Square. When the occupation was denied access to the square, the protestors exposed that what was promised to be public space in planning was actually private and its owners were prepared to strongly assert their claims over this protected property.

The private nature of the land was less explicit in the planning applications for Paternoster Square that were submitted to and approved by the City of London in 1999. The masterplan was prepared by William Whitfield following a previously approved plan led by Terry Farrell (1990). The plans proposed replacing a complex of offices, designed in modern architectural forms and materials, that filled a site of historic streets heavily damaged during World War II. Whitfield's and Farrell's proposals succeeded following other abandoned plans, during the 1980s and 1990s, which had been considered an affront to the historic setting adjacent to St Paul's cathedral. Disagreements over architectural styles and poor financial returns had been the main causes for the contested and abandoned designs for Paternoster Square, which had drawn the attention of architecture critics and conservationists, such as the Prince of Wales. Through the masterplans of Whitfield and Farrell, Paternoster Square was designed following a tradition of landscape which prioritized scenic qualities, individual views and private enclosures. Behind debates about reinstating historic streets, re-establishing vistas and choices of materials, Paternoster Square remained a commercial development which needed to maximize the financial value of the land on which it sat. Mitchell and Staeheli describe a 'practice of property' which includes the 'relations, regimes and struggles over what property is and how it is deployed' (2008, 128). These relations are exerted through exchanges and appropriations of land, and in the case of Paternoster Square, through masterplanning. These practices of property employed landscape as a visual medium to persuade planning officials and future tenants of the value of embracing the development.

Once established outside of the square by St Paul's Cathedral, Occupy LSX grew as an accumulation of large tents for communal purposes, housing the kitchen, bookshop and workshops, and smaller tents in which protestors would sleep overnight. Its forms, representations and actions contrasted with the design and management of the adjacent Paternoster Square development. The unfamiliar makeshift appearance of the public space formed by Occupy LSX represented an embrace of people and issues as well as organizational structures which denied hierarchies of control. After visiting Occupy LSX, Doreen Massey described:

> It was deliberately open, in contrast to the cathedral and the houses of finance all around . . . People came by all the time to listen, to argue. It was a place explicitly of debate.
>
> *(www.publicspace.org, 2013)*

The protestors appropriated the physical space of London to explore and find representation for critical economic and social concerns. Mitchell describes the importance of 'making claims' to public space (Mitchell, 2003, 142). He describes that 'then as now public spaces were only public to the degree that they were *taken* and made public' (2003, 142). Analogously, the privatization of land continually needs to be re-enforced – through regulations, security guards and at Paternoster Square the re-enclosure of what was threatened to be taken back as public space. Through employing traditional landscape techniques for taking over public space, Mitchell writes, 'propertied classes express "possession" of the land, and their control over the social relations within it' (1997, 323). While Occupy LSX finally settled in an in-between space of London to form a new public arena the owners of Paternoster Square gated their public space to reassert their private rights. For several months the sites were inextricably tied together. As privacy and control was tightened and calls for openness and change increased the contrasting conditions of active occupation and securitized enclosure were exacerbated.

Visuals

Cosgrove writes, 'In England, *landscape* popularly denotes an artistic tradition of painting and garden design, historically associated, above all, with the parkland vistas of the eighteenth-century landed estate' (1999, 222). This historic Anglo-Saxon definition of landscapes as visual images and aestheticized spaces opens up contradictions between what is perceived and what is experienced. Harvey describes the 'most important contradiction' of capitalism as 'that between reality and appearance in the world in which we live' (Harvey, 2015, 4). Through the redevelopment of Paternoster Square, its owners and design consultants made claims through drawings to be producing public places alongside which were written promises of 'open public spaces' (Paternoster Associates, 1990). The claims contrast with the assertion of private rights of property through routine patrols of security guards and the confrontation with Occupy LSX protestors who were denied entry. A focus on visual qualities undermines other landscape relations. James Corner explains:

> The pictorial impulse denies deeper modes of existence, interrelationship, and creativity; it conceals the agendas of those who commission and construct it, and it seriously limits the design and planning arts in more critically shaping alternative cultural relationships with the earth.
> *(1999, 158)*

During the months of Occupy LSX the masterplan renderings of a contented public with commercial activities unfolding in the square, drawings which had facilitated planning approval, were contradicted by scenes of less controllable activities of protest and public demonstration. During moments of Occupy LSX ambiguities of what the developers had promised as public space were exposed. What Harvey describes as the 'masks, disguises and distortions' (2015, 4), which had been presented in the masterplan, were removed to reveal a private estate claimed as public space. Expropriations of public spaces by private interests were laid bare, while claims to the provision of public spaces, through which the landowners were afforded additional allowances during the planning process, proved an illusion.

When urban redevelopments are planned in London they are accompanied by picturesque renderings of pristine squares and streets animated with beautiful, diverse and happy people. In *The Dark Side of the Landscape* (1983), John Barrell writes of eighteenth-century landscape paintings:

> by studying the imagery of the paintings, the constraints upon it, and upon its organisation in the picture-space – we may come to see this unity as artifice, as something made out of the actuality of division.
>
> *(1983, 5)*

Attempts by its designers to illustrate the sociability of a redeveloped Paternoster Square through scenic relations belie the editing of images that show comfortable idylls and accessible spaces. Corner writes that architectural works that aim to improve lives are 'reduced under largely representational regimes to simply expressing or commenting on that condition' (1999, 158). Analysing the planning drawings accompanying the Paternoster Square masterplans, we see contemporary urban landscapes adopting scenographic conventions to persuade the viewer of conditions that could not exist and a future that would not unfold. Acceptable social relations, which include welcoming servile doormen, diverse demographics of business people and well-behaved young children, are found across Whitfield's computer-generated renderings and the classically styled watercolour paintings by Edwin Venn for Farrell. Missing from these scenes is the less desirable presence of homeless people, smoking teenagers or political activists, people frequently deterred, and occasionally barred, from London's private estates. Masterplan images of Paternoster Square do not just give a false impression of architectural forms, instead as Barrell claims for eighteenth-century paintings, hidden social agendas are bound up with these idealized scenes. When low-paid workers are depicted, such as doormen to the private buildings, their appearance is made acceptable through their welcoming demeanour. However, and further to Barrell's findings that 'the labouring, the vagrant, and the mendicant poor could be portrayed so as to be an acceptable part of the décor or the drawing rooms of the polite' (1983, 28), undesirable groups are unable to be included in the drawings, and the spaces, of contemporary redevelopments. They present both an unacceptable and uncomfortable presence. Sanitized landscapes are promised in masterplanned scenes, through the removal of contestation, the regulation of unstable social relations and the presence of security personnel. In this way, public spaces as architectural containers in which activities occur are so tightly managed that the potential for public actions is rendered impossible. Critiquing the paintings of Gainsborough, Morland and Constable, Barrell claims that 'the point of the exercise [of the landscape painting] was to suggest that no disjunction existed, and in that way to offer a reassurance that the poor of England were happy'.

Despite the contemporaneity of the computer renderings in Whitfield's proposal both masterplans allude to traditions of landscape painting. The designs of the masterplans established the magnificence of views of the dome of St Paul's cathedral while more functional features of the development are obscured, or as in the case of the ventilation shafts, disguised by Thomas Heatherwick as giant sculptural follies. In contrast to the totalizing masterplan proposals, realized for Paternoster Square and the marketing materials published by its owners, Occupy LSX provided a collage of scenes, ideas, voices and messages. The appropriation of the site adjacent to St Paul's cathedral unfolded by taking and remaking public space rather than fashioning a singular image of protest. Walking through the site, between the common spaces and the private tents in which protestors would sleep overnight contrasted with experiences of architecturally designed public spaces familiar to London. Instead of fountains, lighting and selectively positioned seating, which has become archetypal of well-funded privately led developments, such as *More London*, *Kings Cross Central* and *Paddington Waterside*, a visual aesthetic of low-cost tents and makeshift signs took its place. During Occupy LSX, the veneer of a pictorial landscape, or what Harvey terms the 'surface appearances' (2015, 4), had been removed to present the realities of contested issues and negotiated politics presented in space.

Positions

Traditions of landscape painting present singular perspectives, rendered from static positions of power. As it became defined in England, landscape was a visually and spatially dominated practice, which Cosgrove claims was 'bourgeois, individualistic and related to the exercise of power over space' (1985, 45). It was also 'employed initially to represent spaces of the city' (1985, 49), utilizing linear perspective and privileging the position of the individual viewer. John Berger explains in *Ways of Seeing*: 'Every drawing or painting that used perspective proposed to the spectator that he was the unique centre of the world' (1972, 18). Raymond Williams elaborates, 'For what was being done, by this new class, with new capital, new equipment and new skills to hire, was indeed a disposition of "Nature" to their own point of view' (1973, 123). As landscape painting evolved into the design of landscape space, these ego-centred landscapes conflicted with collective ambitions in and for public spaces. As we have seen in Occupy LSX, and as Bender attests (1993, 2), the way in which landscapes are made and remade opens up multiple and contrasting ways of engaging with our environments. That individuals and groups can simultaneously conceive of and engage with a plurality of landscapes denies static positions and fixed viewpoints – and suggests dynamic potentials of landscapes that are less constrained through narrow visual approaches.

The redevelopment of Paternoster Square was proposed from selected viewpoints, which borrowed scenery, including the dome of St Paul's cathedral, and excluded unresolved architectural spaces and less attractive service features necessary in any urban development. While singular perspectives from the owner's window are not explicitly framed, as Olwig cites of historic landed estates (1993, 332), the views presented in the masterplans are for the benefit of planners who were to be persuaded to approve the proposals and future tenants who were courted to rent office spaces. The control of views was of little concern in Occupy LSX. Instead, non-hierarchical structures of debating and collectively occupying the spaces between St Paul's cathedral and Paternoster Square provided for shared rights and a multiplicity of viewpoints. Attempts by the mainstream media to frame the issues of Occupy LSX (and other Occupy protests around the world) were frustrated by the openness of the debates around issues including the immediate economic crisis, the impacts of austere government policies on education, and unaddressed environmental urgencies. Occupy refused to be defined by a single issue or position. In contrast, the imperatives for economic return generated from the development reflect the singular priorities of their corporate owners, Mitsubishi Estate Co., and their managers Broadgate Estates.

Frames

The gated (or fenced-off) landscape inaccessible within the bounds of Paternoster Square contrasts with the permeable and open spaces of Occupy LSX which found presence in the shadow of St Paul's cathedral. John Tagg explains that the frame 'produces the distinction between the internal and the proper sense and the circumstances' (2009, 246). Over three months the Occupy protesters created alternative forms of public spaces, with their own newspaper, canteen and a bank of ideas. A different order was established across the landscape of Occupy, a presence that was frequently challenged as illegal, with divergent forms of public space from that of Paternoster Square. How Occupy was arranged, what activities were undertaken and many of the complex relations established with local businesses and organizations represented a plurality of ideas, negotiation and debate. Massey describes the openness of the Occupy camp which contrasted with St Paul's cathedral and the highly controlled access to the finance offices in Paternoster Square (www.publicspace.org, 2013). She recalls: 'It was a place of contestation, it was a place of ideas, it was a place of negotiation, and it became, of course, a place which was utterly contested.' The restrictions to Paternoster Square as an open space, as it was filled with an entanglement of steel barriers, and the Occupy protests outside of the square frustrated the scenic intentions of the landowners. On both sides of the buildings that encircled Paternoster Square, the development, which was designed to frame selective views of St Paul's cathedral, was reconstituted through layers of security fences, protestors and enclosure.

Framing of landscapes can be associated with the enclosure of spaces and commodification of land. Cosgrove explains: 'Landscape was, over much of its history, closely bound up with the practical appropriation of space' (1985, 46). Once spaces were taken they could be contained by physical walls, fences and gates and through legal mechanisms of ownership.

> Landscape was framed and reified as a cultural object, to be bought and sold as cultural capital on the burgeoning new art market, much as land itself was being divided up and according to the geometric coordinates of the map, to be sold and traded on the property market.
>
> *(Olwig, 1993, 331)*

The framing of landscape offers a practical means through which specific views are claimed, presented and controlled. The frame affords individuals and organizations that control landscapes an image to edit what is seen, deciding what is included in the landscape and what is excluded from future spaces. As contemporary landscape architects are employed to design public spaces, how they promise new scenes through their drawings can secure potential contracts with their clients and the approval of their projects in the planning process. What designers and their clients choose to include in these landscape scenes, and who is excluded, opens up contradictions in the production of architectural spaces, and especially for developers who claim to provide public access or represent shared concerns. Techniques of framing landscape pictures and space express forms of control – Tagg describes the frame as a 'technology of discipline' (2009, 236) – exerted in the spaces of urban developments through physical barriers, gating and surveillance along with legislations which determine who is permitted access and what they may do once inside. Controls, expressed through physical and legislative frames, aim to control the image of landscapes – visions constructed to enhance the value of privately owned developments and to obscure what Harvey terms, 'the distortions of what is really going on around us' (2015, 4). The frame, as a device, mediates landscape relations between people and land. But not every practice of framing can be associated with individual ambitions for control and commodification. Harvey reminds us that 'not all forms of enclosure can be dismissed as bad by definition' (2012, 70). The enclosure of land for common purposes, or the appropriation of spaces as witnessed during Occupy LSX, highlights the importance of taking back public space while providing clues to the permeability of landscape boundaries. The public space of Occupy LSX suggests that terms of landscape can be inventively reframed through notions of accessibility and engagement rather than exclusion and control.

Post-landscape?

Have we reached a post-landscape condition? Have new designs, representations and physical forms been realized which provide for collective actions and alternative relations with where we live, work and visit? In *Recovering Landscape* (1999), Corner describes his inspiration for advocating a 'recovery' of landscape as 'less the pastoralism of previous landscape formations' but instead the 'yet-to-be disclosed potentials of landscape ideas and practices' (1999, 1). But as economic and political contexts shifted, during the global economic collapse and the subsequent recession, can we identify an emergence of alternative practices and landscape forms? Concerns for ecological restoration and programmatic approaches to landscapes are emphasized by Corner whose firm Field Operations designed the masterplan for New York's Fresh Kills Park (2006) and realized the rehabilitation of the High Line as a public park.[1] However, Corner describes that 'massive process[es] of deindustrialization' have placed new complex demands on land-use planning requiring the 'accommodation of multiple, often irreconcilable conflicts' (1999, 14). Landscape projects that remediate and repurpose polluted post-industrial sites have gained currency in urban redevelopments, building on the work of land artists such as Mel Chin, and landscape architects like Peter Latz.[2] But while we can identify inventive approaches that decontaminate formerly abandoned landscapes, few contemporary landscape or urban design projects have confronted their contribution to increasing land-values, displacement of remaining industries and aggressive gentrification. Environmental recovery of landscapes facilitates urban redevelopment, provides a foundation for spatially and aesthetically reproducing cities and furthers opportunities for economic returns for individuals and organizations that own brownfield sites. Projects improve ecological conditions but fail to address, and in many cases exacerbate, businesses displaced, jobs lost and individuals excluded from renewed urban areas. While in some cases, as Cosgrove claims of recent critical thinking, 'landscape is approached as a spatial, environmental, and social concept rather than as a primarily aesthetic term' (1999, 223), prevailing landscape practices remain tied to economic priorities. And although Corner reminds us that landscape is inextricably 'bound into the marketplace' (1999, 157) neither his writing nor his landscape practice provide clues for how these relations can be uncoupled or rethought.

The contexts of Occupy LSX were specific to a distinct period of economic and political development in London. What occurred as protestors were denied access to Paternoster Square and were forced to occupy the site by St Paul's cathedral was a rare elucidation of the contested landscape of democratic public space. But while the economy may fail or, as Harvey explains, 'stutter and stall and sometimes appear on the verge of collapse' (2015, 11) and while new forms of landscape may appear to emerge in the form of collective actions, the political responses and approaches to London's designed spaces have continued almost unchanged. London's free-market economy, and those of most cities around the world, continue with negligible response. Neither has a post-landscape condition materialized. Projects that succeed in challenging relations between land and economic priorities become increasingly difficult to achieve as private interests and commercial architectural practices dominate urban redevelopments. The construction of new buildings and open spaces accelerated in the years after the 2008 economic crisis, as the spatial commodification of the land on which London resides provided a refuge from less confidently performing stock markets. As property prices increased a literacy of, and discontent with, privatizations of public spaces and the effects of gentrification has widened. Relations between urban development and the displacement of residents from London housing estates and between the design of public spaces and the exclusion of homeless people are increasingly evident to the public. This has led, as Ben Campkin describes in *Remaking London* (2013, 5), to a 'growing scepticism' of the claims made by politicians, developers and architects that they are improving urban areas through development.

Contradictions are exposed as landscape techniques are employed to redesign public spaces. The prioritization of views of public spaces and the adoption of public realm as controlled settings which frame new urban developments frequently denies the lived qualities, the potential for politics and the unpredictable nature of shared spaces. Many new landscapes are claimed by developers and their urban design consultants, to be public places – fulfilling planning obligations through creating attractive settings for their residential and office developments. But Corner explains that the 'veil of pretence that landscape erects is not, however, impermeable' (1999, 157). When these urban redevelopments are completed the presence of poor doors, anti-homeless measures and anti-skate materials, coupled with the enforcement of private regulations, expose the lies of many claims to make public spaces. Mitchell identifies these contradictions, asking 'Landscape or Public Space?' (1997, 322), and claiming that 'aesthetic judgments have the effect of valuing the spaces of the city as landscape rather than public space' (1997, 324). He describes a 'particular definition of "landscape"' (1997, 323) more closely associated with centuries old fixations of landed estate owners who express dominion over spatial, visual and social settings. Citing Lefebvre (1991), Mitchell considers the illusion of landscape which promises both control and comfort.

In contrast, Bender (1993) points to other ways of conceiving of landscapes – ecological and political relations between people and their environments that are difficult to contain or define. These contrasting conceptions of landscapes, as plural entities, appropriated, contested – 'not so much artefact as in process of construction and reconstruction' (Bender, 1993, 3), align more closely with notions of public spaces (see Massey, 2005, 2013), public sphere and publics. Critiquing Habermas (1989), Fraser's description of the public sphere resonates with Bender's assertions for landscape: Fraser states that the public sphere is 'a site of the production and circulation of discourses' (1990, 57) where there is 'a plurality of competing publics' (1990, 61). Similarly, when Massey (2005) describes public space she emphasizes the interrelations and processes which produce a plurality of spaces. Massey claims:

> As well as objecting to the new privatisations and exclusions we might address the question of the social relations which could construct any new, and better, notion of public space.
>
> *(2005, 153)*

Should we therefore consider alternative approaches to landscapes, which are defined by multiple and competing relations? Could new ethics be embraced by landscape architects and urban designers as they are employed in the making and remaking of public spaces? Could temporal frames and spatial bounds be challenged through projects which defy property ownerships and time-frames of urban redevelopment?

If a post-landscape condition is to be sustained relations that address a prevalence of *visual* priorities, static and limited *positions* and narrow, impermeable *frames* will need to be addressed. Alternative approaches to landscape will need to be recovered, invented and practised, reflecting Cosgrove's belief:

> Landscape today is unbounded, flexible, and mobile, composed of forms, connections, and spaces that can neither be contained within conventional frames nor pictured according to the scopic conventions of a distanced, authoring eye.
>
> *(1999, 221)*

If a visual domination of landscape is to be challenged, could landscapes be reconceived through how they work rather than how they look – so that the architectural drawings and constructed spaces more closely accommodate what Mitchell terms the 'messy realities of everyday life' (1997, 323)? How could multiple and collective positions be encouraged in landscape – reflecting expectations for public spaces which represent shared issues and ambitions? And how could we reconceive landscape frames – as permeable, dynamic and shared edges to our images and spaces – which encourages us to look beyond its bounds for what is also happening outside?

Alternatives to scenographic conceptions of landscape could be established while still accepting landscape as a creative medium. Written manifestos have provided means to represent relations between people and land, and which challenge a dominance of visual approaches. In *Manifesto for Maintenance Art* (1969), Mierle Laderman Ukeles challenges distinctions between 'development' and 'maintenance', describes development as 'pure individual creation; the new; change; progress; advance; excitement; flight or fleeing' and maintenance as 'keep the dust off the pure individual creation; preserve the new; sustain the change; protect progress; defend and prolong the advance; renew the excitement; repeat the flight'. She uses the written form of the manifesto to lament the limited opportunity which maintenance allows to explore 'life's dreams' while development offers 'major room for change'. While Ukeles does not explicitly refer to activities of street cleaning, private security or maintaining the condition of public spaces, which are synonymous with private urban developments in London, her manifesto opens up questions of how we people can relate to urban spaces and urban development. Her juxtaposition of the actions of 'development' and 'maintenance' also resonate with leading contemporary landscape architecture practices, such as EMF (Estudi Martí Franch) whose working-through diagrams reconsider landscapes of Girona to embrace incremental phases of maintenance rather than singular transformations of landscapes through masterplans.

Addressing ego-centred approaches and singular positions from which landscapes are viewed through urban redevelopment are challenged if we consider Occupy LSX as a landscape. The non-hierarchical, participatory relations that defined Occupy challenge Lefebvre's critique of landscape as an illusion and in what he describes as the 'delusion of being a participant in such a work' (Lefebvre, 1991, 189). Belief in single architectural authors and trust of individual private landlords could be challenged through understanding the ways through which Occupy emerged and unfolded in space. A written manifesto initiated by the Department of Public Space,[3] and inspired by Michael Sorkin's *Local Code* (1993), adopts social media platform Twitter to encourage students, urbanists and wider publics to collectively author a design code for public spaces in London. The code can be read through

MANIFESTO !

MAINTENANCE ART

Proposal for an exhibition "CARE"

MIERLE LADERMAN UKELES
© 1969

I. IDEAS

A. The Death Instinct and the Life Instinct:

The Death Instinct: separation; individuality; Avant-Garde par excellence; to follow one's own path to death—do your own thing; dynamic change.

The Life Instinct: unification; the eternal return; the perpetuation and MAINTENANCE of the species; survival systems and operations; equilibrium.

B. Two basic systems: Development and Maintenance. The sourball of every revolution: after the revolution, who's going to pick up the garbage on Monday morning?

Development: pure individual creation; the new; change; progress; advance; excitement; flight or fleeing.

Maintenance: keep the dust off the pure individual creation; preserve the new; sustain the change; protect progress; defend and prolong the advance; renew the excitement; repeat the flight;

MIERLE LADERMAN UKELES -2- MAINTENANCE ART

show your work—show it again
keep the contemporaryartmuseum groovy
keep the home fires burning

Development systems are partial feedback systems with major room for change.
Maintenance systems are direct feedback systems with little room for alteration.

C. Maintenance is a drag; it takes all the fucking time (lit.)
The mind boggles and chafes at the boredom.
The culture confers lousy status on maintenance jobs = minimum wages, housewives = no pay.

clean you desk, wash the dishes, clean the floor, wash your clothes, wash your toes, change the baby's diaper, finish the report, correct the typos, mend the fence, keep the customer happy, throw out the stinking garbage, watch out don't put things in your nose, what shall I wear, I have no sox, pay your bills, don't litter, save string, wash your hair, change the sheets, go to the store, I'm out of perfume, say it again—he doesn't understand, seal it again—it leaks, go to work, this art is dusty, clear the table, call him again, flush the toilet, stay young.

D. Art:

Everything I say is Art is Art. Everything I do is Art is Art. "We have no Art, we try to do everything well." (Balinese saying).

Avant-garde art, which claims utter development, is infected by strains of maintenance ideas, maintenance activities, and maintenance materials.
Conceptual & Process art, especially, claim pure development and change, yet employ almost purely maintenance processes.

E. The exhibition of Maintenance Art, "CARE," would zero in on pure maintenance, exhibit it as contemporary art, and yield, by utter opposition, clarity of issues.

MIERLE LADERMAN UKELES -3- MAINTENANCE ART

II. THE MAINTENANCE ART EXHIBITION: *"CARE"*

Three parts: Personal, General, and Earth Maintenance.

A. *Part One: Personal*

I am an artist. I am a woman. I am a wife. I am a mother. (Random order).

I do a hell of a lot of washing, cleaning, cooking, renewing, supporting, preserving, etc. Also, (up to now separately I "do" Art.

Now, I will simply do these maintenance everyday things, and flush them up to consciousness, exhibit them, as Art. I will live in the museum and I customarily do at home with my husband and my baby, for the duration of the exhibition. (Right? or if you don't want me around at night I would come in every day) and do all these things as public Art activities: I will sweep and wax the floors, dust everything, wash the walls (i.e. "floor paintings, dust works, soap-sculpture, wall-paintings") cook, invite people to eat, make agglomerations and dispositions of all functional refuse.

The exhibition area might look "empty" of art, but it will be maintained in full public view.

MY WORKING WILL BE THE WORK

B. *Part Two: General*

Everyone does a hell of a lot of noodling maintenance work. The general part of the exhibition would consist of interviews of two kinds.

1. Previous individual interviews, typed and exhibited.

Interviewees come from, say, 50 different classes and kinds of occupations that run a gamut from maintenance "man," maid, sanitation "man," mail "man," union "man," construction worker, librarian, grocerystore "man," nurse, doctor, teacher,

MIERLE LADERMAN UKELES -4- MAINTENANCE ART

museum director, baseball player, sales"man," child, criminal, bank president, mayor, moviestar, artist, etc., about:"

-what you think maintenance is;
-how you feel about spending whatever parts of your life you spend on maintenance activities;
-what is the relationship between maintenance and freedom;
-what is the relationship between maintenance and life's dreams.

2. Interview Room—for spectators at the Exhibition:

A room of desks and chairs where professional (?) interviewers will interview the spectators at the exhibition along same questions as typed interviews. The responses should be personal.

These interviews are taped and replayed throughout the exhibition area.

C. *Part Three: Earth Maintenance*

Everyday, containers of the following kinds of refuse will be delivered to the Museum:

-the contents of one sanitation truck;

-a container of polluted air;

-a container of polluted Hudson River;

-a container of ravaged land.

Once at the exhibition, each container will be serviced:

purified, de-polluted, rehabilitated, recycled, and conserved

by various technical (and / or pseudo-technical) procedures either by myself or scientists.

These servicing procedures are repeated throughout the duration of the exhibition.

FIGURES 11.3–11.6 Mierle Laderman Ukeles, *Manifesto for Maintenance Art, 1969 'Proposal for an exhibition: "Care"'*. Written in Philadelphia, PA, October 1969. Image credit/permissions: Copyright Mierle Laderman Ukeles. Courtesy of Ronald Feldman Fine Arts, New York.

individual tweets, such as miniature manifesto 'Public Space is where people do their own private things' (www.twitter.com/cuongmay90, 2015). Simultaneously, through selecting specific hashtags, #LondonPublicSpaces and #PublicSpaceCode, collective statements, demands and debates can be read. Challenges to traditional structures of power, such as how ownerships of the vast landed estates are reinforced through their design, management and use, are difficult in the context of prevailing capitalist economies and a politics of prioritizing commercial redevelopment over other public concerns. However, rather than large, comprehensive masterplans which represent ambitions of single landowners' combinations of smaller-scale actions, spaces, buildings and actions could afford a multiplicity of relations, remaking the city as a 'participatory landscape' (Corner, 1999, 159). In contrast to singular perspectival approaches to painting, Berger cites Cubist approaches to collage that represent multiple viewpoints within the same image (1972, 18). Corner elaborates on the potential of '[h]ybridized and composite diagram techniques [which] will allow even further advances in landscape formation because of their inclusive and instrumental capacity' (1999, 166). Again, corresponding more closely with notions of public space, the potential plurality of publics, competing and being contested, suggest how collage approaches could be conceived of in representations and realized in space. Collages, as well as abstractions in the form of diagrams, prove more difficult to contain in static frames and can encourage a questioning of what is not included in the drawing, what is beyond the perimeter of a development or

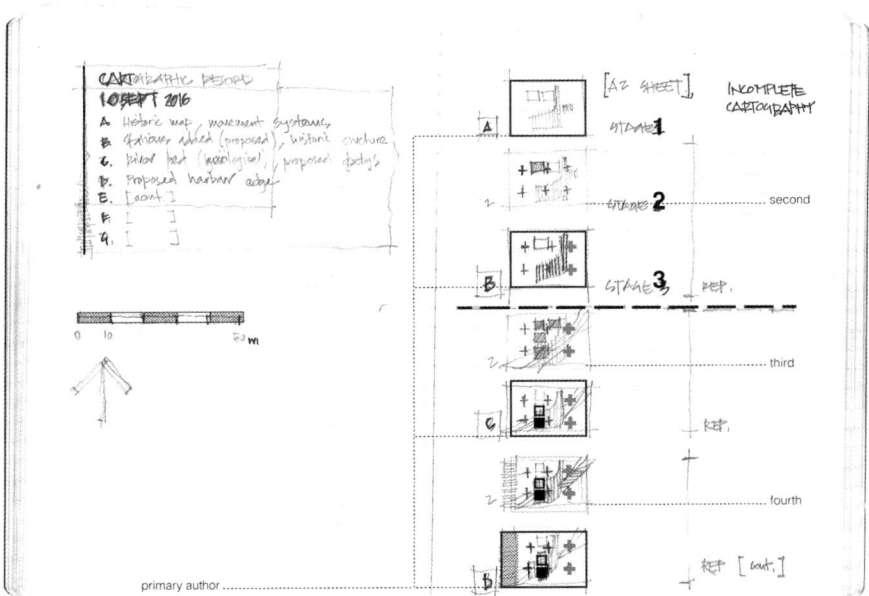

FIGURE 11.7 Co-authored maps, such as Incomplete Cartographies, can entwine contrasting and hidden narratives to establish future design proposals. Image credit/permissions: Ed Wall, 2016.

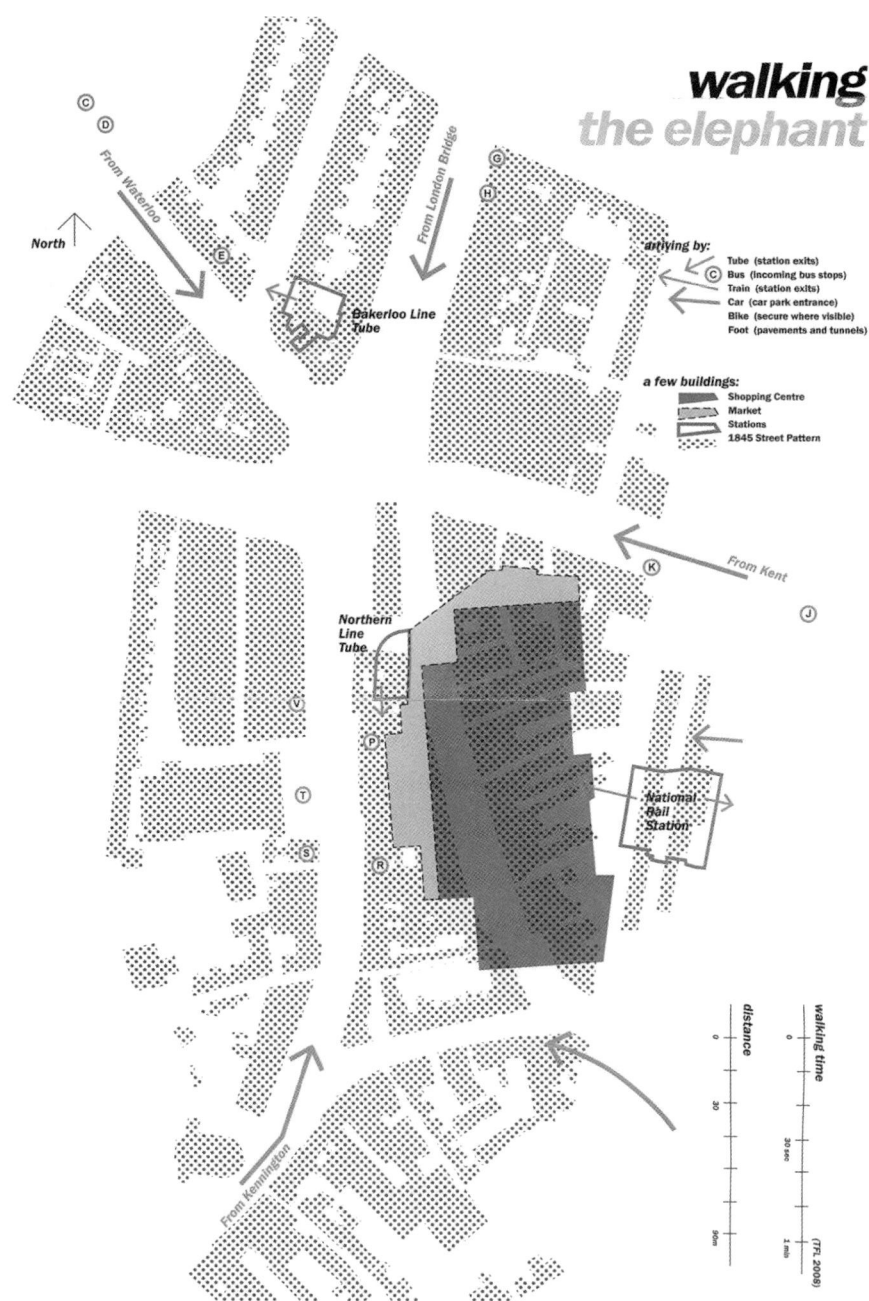

FIGURE 11.8 Research map for *Walking the Elephant* (Dawes and Wall, 2013). Image credit/permissions: Ed Wall, 2013.

Post-landscape **161**

what is not considered in a particular frame of landscape. Hybrid techniques, such as *Incomplete Cartographies* (Wall, 2017), also provide opportunities for collective participation, through more accessible, less professional means of production than architectural drawings. And they blur distinctions between urban design disciplines that propose new spaces and research practices which reflect on past conditions.

Reconsidering frames which have the potential to contain (or open up) and define (or question) landscapes could challenge the legal and physical enclosures of landscapes. *Fake Estates* (1973–1974), by artist Gordon Matta-Clark, is a project of small parcels of land bought in auction from the City of New York. The fragments, which the City had considered untenable, are described by *Cabinet Magazine* (2009) as 'gutterspace' where 'unusably small slivers of land sliced from the city grid through anomalies in

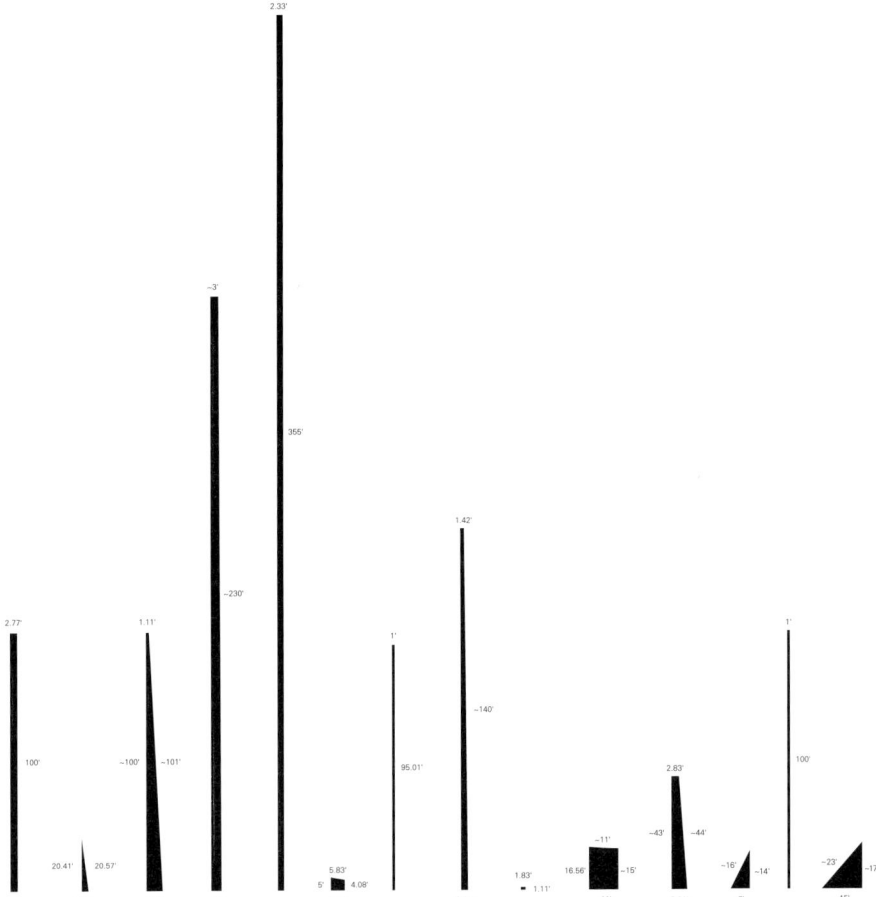

FIGURE 11.9 Odd Lots, an exhibition of Gordon Matta-Clark's *Fake Estates*. Image credit/permissions: Courtesy *Cabinet Magazine*.

surveying, zoning, and public-works expansion'.[4] They were bought by Matta-Clark as he considered them for future artworks and interventions. However, following the artist's death in 1978 the unfulfilled project was archived for several years and the land was returned to the ownership of the City of New York. In 2005 a group of artists, including Ukeles, were invited to participate in a two-venue exhibition, *Odd Lots, Revisiting Gordon Matta-Clark's Fake Estates*. The exhibition aimed to 'de-emphasize[s] the image' of Matta-Clark's work to explore his concerns for 'dematerialization, use value, and systems of social organization'.[5] *Fake Estates* questions issues of land ownership, the exchange of land between public and private interests and it highlights contrasting values of land, from the twenty-five dollars paid by Matta-Clark for each lot, to the negligible use-value that the land had to the municipality. In the context of urban development, projects such as *Fake Estates* contrast with large-scale masterplanning efforts but resonate with the need of developers to accumulate multiple parcels of land, either through agreement with owners or through the employment of legal mechanisms where land can be taken without consent. The project also highlights the legal and political mechanisms, which are bound up with ecological and spatial challenges, which alternative architectural practices need to engage.

Have we reached a post-landscape condition? Although it is difficult to claim that a post-landscape condition has sustained since Occupy LSX, we should be reminded that 'to continue to construe the practice of landscape as the creation of seductive and beautiful images is only to forestall confronting the problems of contemporary life' (Corner, 1999, 158). As if describing the 2008 economic collapse and the public demonstrations that followed, Harvey states: 'Situations arise, however, in which the contradictions become more obvious. They sharpen and then get to the point where the stress between opposing desires feels unbearable' (2015, 2). After 2011, when Occupy protesters claimed new spaces and defined new landscapes in London, discontented voices receded and spatial and visual redevelopment of the city continued apace. As Bender explains that landscapes depend on the temporality of their contexts, whether cultural, social or spatial, we could expect a fleeting ephemerality of landscapes which are continually contested and remade. At moments such as the years after the economic collapse and the evictions of Occupy protesters from sites across London, Harvey claims: 'The genie is, as it were, temporarily stuffed back into the bottle' (2015, 3). However, new conditions will undoubtedly be established, challenging visually formulated relationships with the land and creatively disrupting other narrowly perceived and overtly controlled landscapes of our cities.

Notes

1 The first phase of the High Line opened in 2009 with the final section due to open in 2017.
2 See Mel Chin's Revival Field (1991) and Latz and Partners Duisburg Nord Landscape Park (1990–2002).
3 https://twitter.com/deptpublicspace, 2016.
4 www.cabinetmagazine.org/events/oddlots.php, 2016.
5 www.cabinetmagazine.org/events/oddlots.php, 2016.

References

Barrell, J. (1983) *The Dark Side of Landscape*. Cambridge: Cambridge University Press.
Bender, B. (1993) *Landscape: Politics and Perspectives*. Ann Arbor, MI: University of Michigan Press.
Berger, J. (1972) *Ways of Seeing*. London: Penguin.
Campkin, B. (2013) *Remaking London*. London: IB Taurus.
Corner, J. (ed.) (1999) *Recovering Landscape: Essays in Contemporary Landscape Architecture*. New York: Princeton Architectural Press.
Cosgrove, D. (1985) 'Prospect, Perspective and the Evolution of the Landscape Idea', *Transactions of the Institute of British Geographers* 10(1), pp. 45–62.
Cosgrove, D. (1999) 'Airport/Landscape', in J. Corner (ed.) *Recovering Landscape: Essays in Contemporary Landscape Architecture*. New York: Princeton Architectural Press.
Dawes, A. and Wall, E. (2013) 'Cartographies and Itineraries: Walking through the Elephant and Castle Market', NYLON Conference 2013, London.
Fisher, M. (2009) *Capitalist Realism: Is There No Alternative?* Alresford, Hants: 0 Books.
Fraser, N. (1990) 'Rethinking the Public Sphere: A Contribution to the Critique of Actually Existing Democracy', *Social Text* 25/26, pp. 56–80.
Habermas, J. (1989) *The Structural Transformation of the Public Sphere: An Inquiry into a Category of Bourgeois Society*. Cambridge: Polity Press.
Harvey, D. (2012) *Rebel Cities: From the Right to the City to the Urban Revolution*. London: Verso.
Harvey, D. (2015) *Seventeen Contradictions and the End of Capitalism*. New York: Oxford University Press.
Hunt, J.D. (1994) *Gardens and the Picturesque: Studies in the History of Landscape Architecture*. Cambridge, MA: MIT Press.
Lefebvre, H. (1991) *The Production of Space* (trans.). Oxford: Basil Blackwell.
Mason, P. (2015) *PostCapitalism: A Guide to Our Future*. New York: Farrar, Straus and Giroux.
Massey, D. (2005) *For Space*. London: SAGE.
Massey, D. (2013) 'Space and Power are Intimately Related: Interview with Doreen Massey', online available at: www.publicspace.org/en/post/space-and-power-are-intimately-related (accessed 13 January 2017).
Mitchell, D. (1997) 'The Annihilation of Space by Law: The Roots and Implications of Anti-homeless Laws in the United States', *Antipode* 29(3), pp. 303–335.
Mitchell, D. (2003) *Right to the City: Social Justice and the Fight for Public Space*. New York: The Guilford Press.
Mitchell, D. and Staeheli, L. (2008) *The People's Property? Power, Politics and the Public*. Abingdon: Routledge.
Olwig, K. (1993) 'Gods and Humans', in B. Bender (ed.) *Landscape: Politics and Perspectives*. Ann Arbor, MI: University of Michigan Press.
Paternoster Associates (1990) *Paternoster Square: The Masterplan*. London: The Associates.
Sorkin, M. (1993) *Local Code*. New York: Princeton Architectural Press.
Tagg, J. (2009) *The Disciplinary Frame: Photographic Truths and the Capture of Meaning*. Minneapolis, MN: University of Minnesota Press.
Ukeles, Mierle Laderman (1969) *Manifesto for Maintenance Art, 1969! Proposal for an Exhibition 'Care'*. New York: Ronald Feldman Fine Arts, Inc. (accessed 18 April 2007).
Wall, E. (2017) 'Incomplete Cartographies: A Methodology for Unfinished Landscapes', in *OASE #98 Narrating Urban Landscapes*. New York: NAi Publishers.
Williams, R. (1973) *Country and the City*. New York: Oxford University Press.

12
ACTIVATING EQUITABLE LANDSCAPES AND CRITICAL DESIGN ASSEMBLAGES IN BANGKOK

Camillo Boano and William Hunter

Introduction

Landscape is at once a contentious, anxious, and opportunistic term – always contingent on the action of making and re-making. And it is within our actions and reactions to landscape that *agency* manifests itself. This is certainly true among the prototypical rapid urbanism of so-called developing arenas, where the fabric of Bangkok, for example, has emerged as an inimitable testing ground for the barriers of space and scale in bridging formal and informal landscapes. Within Bangkok's maze of busy canals, roads and overhead trains, the Community Organizations Development Institute (CODI)'s *Baan Mankong* housing and community upgrading programme confronts the shifting spatiality of the city and the obstacles to multi-scalar community intervention. Further embedding the issues of informal settlement upgrading and infrastructural development into the underlying institutional frameworks, the programme seeks to cultivate a ground where the previously excluded can take charge of their fate.

By opening a conceptual gate for a new model of creative practice, territorial organization and the production of equitable landscapes, the *Baan Mankong* activates the notion of landscape as a realm of manoeuvre where both design knowledge and space are reconfigured in a tactical manner. For such critical agency to be sharpened however, landscapes must result in a new spatiality that does not merely maintain the stasis of homogenized upgrading, but rather capitalizes on responsive design outcomes that reframe the canvas from which they are composed.

Speaking of *landscape* in this sense is essentially a de-facto way of speaking of us and our *agency*. Yet, speaking of landscape also of course alludes to a kind of territory – an objectification of space (Elden, 2013; Masiero et al., 2015). If we consider space as driving us to the abstract, the limitless, landscape forces us towards materiality, to think on borders, lines and *topoi*, as well as the forms of life that belong to it. Landscape is space that takes shape and at once becomes land and terrain. For us, and especially in the case of Bangkok that we will present here, landscapes are very fluid in both the physical sense and in the sense of action towards them. The approach is a more positive and opportunistic stance than say Creswell's view that landscape 'does not have much space for temporality, for movement and flux, and mundane practice' (Cresswell, 2003). While we agree with his assertion and will highlight later that landscape practice must always be critical in nature, we do not necessarily perceive landscape itself as being something fixed or already accomplished. Hence the complicated process of enacting upon and re-acting to the evolving changes, everyday stasis and surprise occurrences brought to our landscapes. Recognizing these divergent viewpoints signal a myriad of ways in which landscape is being categorized, utilized and transformed into not only the ground in which we work, but the manner in which we construct it. The previous chapters in this book have alluded to this spectrum in one way or another, from the emerging epistemology of landscape urbanism to rights of power, thus confirming the depth in which landscape is considered and applied in research and practice.

In this chapter, we engage initially with the historically unique ecological landscapes of Bangkok and align our charge in the present day from the localized perspective of semi-government housing programmes and community development. In Bangkok, as in other contested arenas, the idea of *agency* has grown around a movement of (in)equality and accessibility that is not only important amidst the processes of city (re)making, but essential to any underlying participatory action. It has provided local practitioners and communities alike with a purpose and direction to assert alternative methods and solutions that are outward looking, engaging and transformative (Hilal et al., 2014; Bradley, 2015; Douglas, 2015; Grubbauer, 2015; Astolfo et al., 2015). Such a nuanced and complex consideration of landscape and the social and political veins of engagement evokes the theoretical social justice positions of Soja, Rawls and others as we adopt a particular reading of Jacques Rancière's views on equality, thus directing the agency of landscape toward that of a protagonist driven by transformative channels and status quo critique.

Posing a Rancierian critique of political aesthetics and equality alongside the design actions of community architects and everyday citizens in Bangkok, we expose the notion of *agency* as having the ultimate potential for new spatial interpretations of the *landscape*. In doing so, *landscape* is envisioned not merely as a type of practice or realm upon which design is enacted, but rather as the ether through which all life-affirming socio-political activity unfolds.

A very brief evolution of Bangkok's urban landscapes

To understand the potential for local agency within Bangkok's complex territorial landscapes, one must realize its distinctive history and its increasingly urbanized reality. Thailand's urbanism has evolved from a tributary political system and a distributary water system that followed the Chao Phraya River from upstream to downstream. Bangkok is located downstream from the previous capitals of Sukhothai and Ayutthaya (McGrath et al., 2013). While north–south *khlongs* (canals)

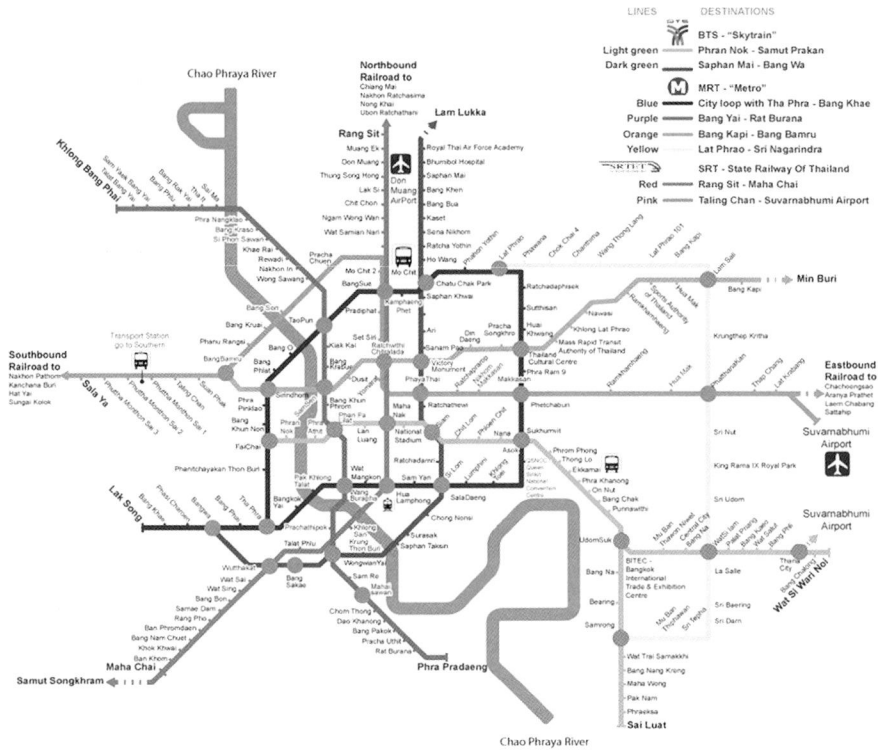

FIGURE 12.1 Bangkok Transit Map with Chao Phraya River (transitbangkok.com/property-D)

FIGURE 12.2 Bang Bua Thong District *khlong* (canal) and community with repurposed canal-side pathway connection. Image: William Hunter.

were formed to enhance trade, shortening the distance for boats travelling from the Gulf of Thailand to Ayutthaya, and later to Bangkok, other east–west *khlongs* were created for defence purposes and to allow troop movements. The Bangkok Noi Canal, one of the busiest in the city, is a historical course of the Cho Phraya River. In one emerging urban condition, where the Bang Wa Skytrain interchange exists, a new fast boat dock is proposed to exit directly into the Skytrain platforms which would begin to formally integrate these two distant eras of transit.

The periphery of Bangkok is changing fast as well, where surrounding provinces are connected by a new ring road and the expanded Skytrain and highways. The example above of modal juxtaposition between the Skytrain and fast boats exists within a broader, lengthier transition from the historically slow water-based city to the fast car-based city where fluid *khlongs* have given way to clogged roads (Sintusingha, 2010). The long corridors of Bangkok's historic *khlongs* created a type of linear urbanism that saw canals often filled in to create roads, or roads being built parallel to canals. *Thanons* (roads) ignored pre-existing land patterns and were built to link distant provinces.

Since the 1950s, rural villages adjacent to new roads have developed as part of the car-based city. *Soi* (alleys or local feeders) are built to connect to old water-based temples and villages as well as to create land development estates or housing development estates. The landowners built *soi* that follow agrarian land ownership patterns, therefore each landowner disparately participates in piecemeal and unregulated development extending into Bangkok's suburbs (Sintusingha, 2006).

The recent construction of the ring road and the new airport has increasingly connected these maze-like radial corridors at the periphery. In many cases, informal settlements and villages have sprouted along these pathways along the periphery, as well as within central Bangkok. And despite the increasing connections by road and canal, these settlements remain largely underserved, unregulated and vulnerable. Their true access to the city including common resources and sufficient housing is still difficult to achieve. Given increasing urbanization throughout the world and the proliferation of informal settlements in many cities, it is important to take a closer look at how Bangkok and certain organizations like CODI have made progress in upgrading and improving living conditions in these settlements (Bhatkal and Lucci, 2015).

CODI and the *Baan Mankong*

Due to a historical lack of sufficient and affordable housing in Bangkok, communities have settled among the cracks in landscape. This could be diagnosed as social and institutional 'pocket-urbanism', a situation that has formed barriers of interaction among communities and certainly between communities and authority figures. Furthermore, the decision-making processes revolving around government housing policies also manifest in a 'pocket-like' manner, having missed the links between planning, design and the true needs of communities they are deemed to represent. Here, the power of landscape seems applied rather than shared, further contributing towards a vicious circle of pocket-segregation.

FIGURE 12.3 Informal settlement communities dot the landscape in juxtaposition to new luxury residential towers. Image: Camillo Boano.

As a regional node of operation within the widely cast net of the Asian Coalition for Housing Rights (ACHR) in South-East Asia, CODI has led the charge for grassroots agency within the Bangkok metro region. In supposed parallel opposite to the contract-driven, pre-designed housing of the National Housing Authority's Baan Eua Arthorn Programme, *Baan Mankong*, under CODI's leadership, stresses the participation and initiative of people as the central tools in scaling up the programme to the national level. Local people and community organizations are intended to control the funding and management and thus, in many cases, certain flexible finance schemes have allowed for the planning and implementation of 'tailored' building and landscape intervention projects that address specific community needs, priorities and aspirations.

To a large degree, the work of CODI and particularly the *Baan Mankong* programme has set a new agenda towards alleviating the misrepresentation and exclusion of communities. If these occurrences have led to a kind of 'pocket urbanism' where communities are cut-off or marooned, it could be said that CODI has achieved a level of 'de-pocketizing' the city's landscape through its integration of geographically disparate communities and individuals around a commonly communicated goal. The organization has envisioned housing policies beyond the simple provision of shelter, putting people at the centre of the process of community-wide regeneration. Through their attempts at empowering the poor by building social networks, they open up a negotiating flexibility with government authorities that can improve housing processes and conditions at the spatial level.

FIGURE 12.4 A CODI-funded *Baan Mankong* project rises from the ground. Image: William Hunter.

Baan Mankong aims at cultivating an urban landscape wherein people who have previously been excluded from secure housing can take over the lead in the process of their own secure housing provision. 'It is premised that the people in need have a massive potential and are thus more resilient to take their housing, environmental and territorial issues into their own hands' (Boonyabancha et al., 2012, 433). Between 2004 and 2014, the *Baan Mankong* programme provided tenure security to over 96,000 households (CODI, 2014). Security of tenure rose from 88 per cent of the urban population in 1990 to 95 per cent in 2010 (National Statistical Office, 2010), remarkable progress having occurred amidst increasing urbanization (Bhatkal and Lucci, 2015).

In *Baan Mankong* physical change is conceived and practised as a vehicle for social change. Physical upgrading of informal areas, from landscapes to houses, is thus a twofold function of improving the material reality of the urban poor while fostering confidence in historically marginalized groups concerning their skills and capacities, individually and collectively. This calls to mind the interpretive landscapes of Duncan as 'the central element in a cultural system – for as an ordered assemblage of objects, it acts as a signifying system through which a social system is communicated, reproduced, experienced, and explored' (1990, 17). Such visible agency suggests and encourages those communities in similar situations to follow alternative possibilities in order to expand the resilience and transformative potential of the whole system. This is an iterative process in which, over time, material improvements reinforce the terms of engagement with different actors and vice versa, building up strength and power in and of the communities. Boonyabancha and Mitlin (2012, 403) explain that this ambition, working across scales, has 'two underlying dimensions: first, the creation of institutions based on relations of reciprocity [within communities]; and second, the strengthening of relations between low-income community organizations such that they can create a synergy with the state' while on a national-scale push for policy change and wider political recognition giving low-income communities agency in their own territorial creation.

The emerging agency of the community architect in Thailand

In regards to the notion of increased attention and *agency*, a key component in the CODI umbrella is the emerging existence of what is referred to as the 'Community Architect'. The concept of 'community architects' is highly respected, clearly paramount and widespread throughout Bangkok and the *Bang Mankong* programme. This growing clan was initially and still is composed of young architects, landscape designers and planners trained at local universities in the Bangkok region who chose to eschew the typical architectural career path. The identity nomenclature and agenda of these individuals has expanded to all those who may facilitate housing processes as well as those sitting in parallel to a clustering of anti-mainstream architectural voices in South-East Asia. This loosening of the term in some instances has opened the door for genuine levels of criticism. Through the research of University College London's Development Planning Unit[1] and others, major questions have arisen in regards to the instrument and role of 'community architect' in the design and facilitation of housing initiatives. In turn, their tasks, expertise and overall involvement can be debated as to the degree in which 'community architects' are actually helping communities evolve or simply masking genuine effort with self-indulgent action and appearances.

While among the urban landscapes of Bangkok it has largely been concluded that these practices have contributed significantly to the international 'ethical turn' in architecture (Boano and Kelling, 2013; Boano, 2014; Bradley, 2015; Douglas, 2015), design practitioners are still searching for a specific role in investigating the complexities of architecture and urbanism. And they yearn towards understanding their professional and disciplinary shift in now, more than before, shaping spaces that engage communities and enable social justice. On the other hand, and in addition to social turns, development practitioners have gone through a reflexive rediscovery of how architecture, design and landscape may be experimenting with a new pragmatic radicalism oriented towards more critically sophisticated outcomes. It remains to be seen whether these professional and disciplinary shifts are genuine transformations or the work of neoliberal jargon and marketing trends. Surely the challenges of the humble are expounded by those of the appropriator every day. But it is also a responsibility of the genuine element to maintain its course.

Acknowledging that true participatory design requires additional skill, values and creativity from individuals beyond those practised in conventional practice, the question of how to expand this agency within the mainstream disciplines emerges as a significant purpose of CODI and the various professionals working within the *Baan Mankong* programme. 'Community architects are trained to think beyond the conventional concerns of building design. It is in the nature of their profession to

FIGURE 12.5 Community architects, students and residents collaborate over future development plans and strategies. Image: William Hunter.

"design" – to transform from "what is" into "what it could become"' (Luansang et al., 2012, 502). The practitioner here assumes the role of translating design inputs from their own people and shows how they have been 'the key essence of the process. If a community architect can help explain that transformation process properly to the larger society, it becomes a kind of empowerment' (ibid.).

The agency in Rancière's political aesthetics

If in fact the case study of CODI and the *Baan Mankong* revolves around the notion of equality and self-action, the political views of Jacques Rancière particularly offer us much to reflect upon and utilize. Despite never directly indulging in defining landscape, Rancière's political theory revolves around aesthetics that for him is first and foremost the sphere that conditions the situated and historical configuration of social space. As such, aesthetics operates on an ontological level as the principle for the spatial ordering and social dealing with 'the sensible' – in Rancière's context understood as that which concerns the senses and which is played out on a shared and material level (Rancière, 2009). As such, aesthetics and its agency could be defined as an inter-subjective framework, regulating the relationships between subjects and surroundings. Therefore, landscapes share sensibilities upon which subjectivities, commonalities and realities may be played out. What we need to accept, however, is that the invention of landscape is 'always ambiguous, precarious, litigious' (Rancière, 2010, 202) and for him the essence of politics is 'the break with the "normal" distribution of positions' in such landscapes and the continuous manifestation of a rejected or vanishing diversity (Rancière, 2010, 27–44). For Rancière, politics is always aesthetic in this basic sense, as an actualization of different ways of sharing sensible landscapes. Politics can also never be static and pure as it is characterized in terms of division, conflict and polemics that allow the invention of the new, the unauthorized and the disordered, the unexpected.

Locating democracy at the centre of his political theory, our aim and interest in Rancière here is the reversal strategy that he is staging, where instead of putting equality as a result at the end of the process, he asserts to put it at the beginning. The relevance of Rancière's theorization within the contested and challenging landscapes of Bangkok is that it allows for a material, sensorial and concrete formulation of politics, political participation and enactment that are based on a different vision of equality as a reconfiguration of a space 'where parties, parts or lack of parts have been defined . . . making visible what had no business being seen, and makes heard a discourse where once there was only place for noise' (Ranciére, 1999, 30).

In Bangkok, it could be considered that real politics exists only 'when the natural order of domination is interrupted by the institution of a part of those who have no part' (Ranciére, 1999, 11) and 'there is the appearance of a subject, the people' (ibid., 86). As such the irruption of new *political* bodies who commit to managing themselves and their spaces autonomously – continuously struggling to embody and stage equality and finding alternative way of framing resilience, is a central landscape in which to discuss what he calls 'specific scenes of contradictions'. Such an active and perpetual gesture is then analysed as the central contribution in the realm between justice and resilience.

This echoes the *spatial justice* charge of Ed Soja and John Rawls whereby space is not simply an indicator of justice but a commodity to be distributed (Williams, 2013). And it certainly calls to mind Harvey's clamouring on people's struggle for a *right to the city* as 'far more than the individual liberty to access urban resources', but rather 'the right to change ourselves by changing the city' (Harvey, 2008). Similarly, for Rancière a political struggle occurs when the excluded seek to establish their identity, by speaking for themselves and striving to get their voices heard and recognized as legitimate – disrupting the specific horizon and modalities of sensory experience. Such struggle is evident in many of Bangkok's marginalized communities struggling for space, resources and nature. Here, people have leveraged collective resources as bargaining power to claim politically legitimate participation in the development of landscapes and thus reclaiming their clear position in the resilience–justice continuum. These themes of equality attribute the agency of change to the excluded residents now in the *Baan Mankong* programme and therewith enact a radical break with conventional participatory development practice and the environmental justice debate that is both distribution-focused and recognition-centred.

Awakening people's energy and voices

Urban poor groups and other grassroots organizations are fundamental components of the production of the whole city: they are the ones who keep it going. This is crucial, since it also sparks political responsibility and a sense of ownership over the process and the results it will produce. Design comes from many actors, though mainly from the urban poor groups themselves, and is reconfigured not simply as another statement (although maybe a 'participatory' one) in the overall functioning of an already existing governmental structure, but, rather, functioning as theatrical manifestation of the people's emancipatory potential through city-wide action. *Baan Mankong* participants are connected with diverse actors such as local authorities, servicemen, landowners, as well as NGOs and academia: 'instead of the city being a vertical unit of control, these smaller units – people-based and local – can be a system of self-control for a more creative, more meaningful development' (Boonyabancha, 2005, 22–23). Contrary to conventional strategies of simply providing physical assets – where housing is treated as a technical result rather than a political issue – the programme's ambition goes beyond generating power on the side of historically marginalized people through collective organization so that they become *legitimate development subjects* (Boano and Kelling, 2013).

CODI and *Baan Mankong* have long recognized that affording appropriate recognition to marginalized and vulnerable groups within urban policy and planning is fundamental to identifying where the need to build resilience

FIGURE 12.6 A maturing CODI community exudes ownership, pride and resilience against a challenging marsh landscape. Image: Camillo Boano.

is greatest. Both resilience and justice depend on recognizing the plurality of knowledge types and of governance systems used around the world to manage risks (Adger, 2006). What clearly emerges in the presupposition of inclusion as central to the *Baan Mankong* programme is a critique of numerical teleology. The underlying logic of awakening people's energy is thus achieved through a strategic reconfiguration of the political landscape. This considers existing identities and subjects and presupposes their equality, which drastically changes the status quo of individuals and communities who are not only invited to participate (numerical) but have power and agency, and in this case housing and resources redistributed in their favour.

In refocusing attention on equity, environmental justice has the potential to provide productive intellectual and political landscapes. These constitute multidimensional and multi-scalar explorations of their many meanings, manifestations and implications (Walker and Bulkeley, 2006).

Forging ahead through unpredictable landscapes

CODI faces vast challenges through its *Baan Mankong* housing and community upgrading programme in confronting the spatiality of the city and its obstacles to multi-scalar intervention. However, it has certainly paved the way for new modes of practice, community integrated organization, and the production of alternative landscapes that are more equal and accessible. As an assemblage of disparate and dispersed communities with varying degrees of cohesion and mobilization, the *Baan Mankong* programme further embeds the issues of informal settlement upgrading and infrastructural development into the underlying institutional framework that they all share. The ability for this critical design assemblage to transcend the barriers of space (housing) and scale (informal landscapes) hinges on the depth and potency of institutional roots in their growth towards city-wide solutions.

The learning from the Bangkok case illustrated herein can be expanded in a general capacity of our sensitivity that results from an activity of partition and of partaking. CODI's experience invites us to see ourselves as actors, and to trade the vocabulary of political acts normally shaped around actions and reactions, activity or passivity, for a vocabulary of political gestures. The sphere of politics thus appears as a theatrical stage rather than as a battlefield. Ultimately, such a reconfiguration offers to reveal the lines of power and agency that are written and rewritten in urban landscapes, and to contest the spatial ordering that assigns everyone and everything its proper place.

Note

1 This chapter is based on research collaboration between the Development Planning Unit (DPU) and the Community Organizations Development Institute (CODI), the Asian Coalition for Housing Rights (ACHR) and the Community Architect Networks (CAN). In particular, it is the result of reflections on the *Baan Mankong* Housing Programme

that have emerged in the course of three research projects by two of the DPU's Masters programmes (MSc Building and Urban Design in Development and MSc Urban Development Planning), which took place in 2011, 2012 and 2013 in Bangkok, where several communities at different stages of implementation were involved.

References

Adgar, W.N. (2006) 'Resilience, Vulnerability, and Adaptation: A Cross-Cutting Theme of the International Human Dimensions Programme on Global Environmental Change', *Global Environmental Change* 16(3), pp. 268–281.

Astolfo, G., Talocci, G. and Boano, C. (2015) 'A Six-fold Mandate for an Engaged Urban Design Research Education', *Urban Pamphleteer* 5, pp. 43–45, www.ucl.ac.uk/urbanlab/research/urban-pamphleteer/UrbanPamphleteer-5.

Bhatkal, T. and Lucci, P. (2015) 'Community Driven Development in the Slums: Thailand's Experience', ODI, London. Available at: www.odi.org/sites/odi.org.uk/files/odi-assets/publications-opinion-files/9668.pdf.

Boano, C. (2014) 'Architecture of Engagement: Informal Urbanism and Design Ethics', *Atlantis Magazine* 24(4), pp. 24–28.

Boano, C. and Kelling, E. (2013) 'Towards an Architecture of Dissensus: Participatory Urbanism in South-East Asia', *Footprint* 7(2), pp. 41–62.

Boonyabancha, S. (2005) 'Unlocking People Energy', *Our Planet: The Magazine of the United Nations Environment Programme* 16(1), pp. 22–23.

Boonyabancha, S. and Mitlin, D. (2012) 'Urban Poverty Reduction: Learning by Doing in Asia', *Environment and Urbanization* 24(2), pp. 403–421.

Boonyabancha, S., Carcellar, F.N. and Kerr, T. (2012) 'How Poor Communities Are Paving Their Own Pathways to Freedom', *Environment and Urbanization* 24(2), pp. 441–462.

Bradley, K. (2015) 'Open-Source Urbanism: Creating, Multiplying and Managing Urban Commons', *Footprint* 16 (Commoning as Differentiated Publicness), pp. 91–108.

CODI (2014) 'Progress Report of Baan Mankong', September, Community Organizations Development Institute, Bangkok.

Cresswell, T. (2003) 'Landscape and the Obliteration of Practice', in K. Anderson, M. Domosh, S. Pile and N. Thrift (eds) *Handbook of Cultural Geography*. London: SAGE, pp. 269–281.

Douglas, G.C.C. (2015) 'The Formalities of Informal Improvement: Technical and Scholarly Knowledge at Work in Do-it-yourself Urban Design', *Journal of Urbanism: International Research on Placemaking and Urban Sustainability*. Published online 9 April 2015. doi: 10.1080/17549175.2015.1029508.

Duncan, J. (1990) *The City as Text: The Politics of Landscape Interpretation in the Kandyan Kingdom*. Cambridge: Cambridge University Press.

Elden, S. (2013) *The Birth of the Territory*. Chicago, IL: Chicago University Press.

Grubbauer, M. (2015) 'Not Everything is New in DIY. Home Remodelling by Amateurs as Urban Practice', *Ephemera* 15(1), pp. 141–162.

Harvey, D. (2008) 'The Right to the City', *The New Left Review* 53 (accessed 4 April 2016).

Hilal, S., Petti, A. and Weizman E. (2014) *Decolonizing Architecture, Architecture after Revolution*. Berlin: Sternberg Press.

Luansang, C., Boonmahathanakorn, S. and Domingo-Price, M.L. (2012) 'The Role of Community Architects in Upgrading: Reflecting on the Experience in Asia', *Environment and Urbanization* 24(2), pp. 497–512.

Masiero, R., Assennat, A. and Longo, A. (2015) *Paesaggio Paesaggi. Vedere le Cose*. Foggia: Libria.

McGrath, B., Tachakitkachorn, T. and Thaitakoo, D. (2013) 'Bangkok's Distributary Waterscape Urbanism', in Kelly Shannon, Bruno De Meulder and Yanliu Lin (eds)

Village in the City: Asian Variations of Urbanisms of Inclusion. Chicago, IL: Park Books – UFO: Explorations of Urbanism.
National Statistical Office (2010) 'Population and Housing Census', National Statistical Office, Bangkok.
Rancière, J. (1999) *Disagreement: Politics and Philosophy*. Minneapolis, MN: University of Minnesota Press.
Rancière, J. (2009) *The Emancipated Spectator*, trans. Gregory Elliott. London: Verso.
Rancière, J. (2010) *Dissensus: On Politics and Aesthetics*, trans. Steven Corcoran. London: Continuum.
Sintusingha, S. (2006) 'Sustainability and Urban Sprawl: Alternative Scenarios for a Bangkok Superblock', *Urban Design International* 11, pp. 151–172.
Sintusingha, S. (2010) 'Bangkok's Urban Evolution: Challenges and Opportunities for Urban Sustainability', in A. Sorensen and J. Okata (eds) *Megacities: Urban Form, Governance, and Sustainability*. Tokyo: Springer.
Walker, Gordon and Bulkeley, Harriet (2006) 'Geographies of Environmental Justice', *Geoforum* 37(5), pp. 655–659.
Williams, J. (2013) 'Toward a Theory of Spatial Justice', 'Theorizing Green Urban Communities' Panel, Annual Meeting of the Western Political Science Association, Los Angeles, CA, 28 March.

13
AGENCY AND ARTIFICE IN THE ENVIRONMENT OF NEOLIBERALISM

Douglas Spencer

Introduction: the emergence of 'emergence'

Landscape urbanism, as Charles Waldheim remarks in his recent *Landscape as Urbanism*, has 'emerged' of late as a paradigm for engaging with the design of the contemporary city.[1] The term 'emerged' is, in this context, an especially loaded one. In the discourses of ecology, complexity and self-organization – within which much of what counts as landscape urbanism has been embedded – 'emergence' suggests a coming into being absent of external organization by human agency, artifice or intervention. What 'emerges' is said to be of an order of complexity and intelligence beyond that of its component parts. A swarm of insects emerges from the immediate coordination of the unconscious behavioural patterns of the individual insects of which it is composed. It achieves a consistency of being without recourse to any individually superior intelligence. In describing a model of urban design as having *emerged* there is, however, something more at stake. Landscape urbanism has not, after all, conjured itself into existence, but has been consciously designated, formulated and theorised as such. The significance of this extends beyond landscape urbanism, and is indicative of a broader tendency within architecture, landscape and urban design towards the disavowal of conscious agency that is captured in the term 'emergence'; a designation that tends to obscure its own deployment alongside the conceptual labour it is put to in positing creation absent of conscious design. Designers distance themselves from their own creations – whether these be designs or models of design – preferring to locate the origins of these in the natural unfolding of being. In the process their own artifice is concealed. By 'artifice' I intend an 'art of making'; a practice of fabrication conducted for and from within a set of social, material, economic, political, cultural and technological conditions. Artifice is, in this sense, no less real than anything that might be described as 'emergent'. The realities of the artificial are, however, made opaque to critical reflection – as the work of socially, politically and economically interested parties – in being presented as attributable to the work of purely disinterested agents: nature, emergence, self-organization, the way of the world.

What follows is, in part, a genealogy of the disavowal of agency and artifice on the part of the designer. In this, I attempt to show how the effacement of agency – especially as exercised through the mobilization of discourses of emergence, ecology and environment – has served to obscure the ways in which landscape design is implicated within forms of biopolitical regulation. From the regional scale of Ebenezer Howard's Garden City model, and its early and mid-twentieth-century development, through to the promotions of a global and cosmic environmental consciousness within design, and up to its current manifestations such as landscape urbanism, I also chart a progressive abandonment of critical and overtly political agendas. This trajectory, especially in its identification with so-called bottom-up and environmental processes, has now, I argue, effectively brought design into alignment with the orbit and agenda of neoliberalism.

Biopolitical regulation: from imperial ecology to the *stadtlandschaft*

If power today can be understood in terms of biopolitical regulation, of a productive management of the relations 'between the organic and the biological, between body and population', then ecology and environment surely count among its most vital instruments.[2] The science of ecology owes its origins, after all, to the nineteenth-century imperialist projects in which it was employed to orchestrate and control the productivity of species and subjects captured within colonial dominion. As Peder Ankar argues, in his *Imperial Ecology*, the new science served the British Empire with the 'tools for understanding human relations to nature and society in order to set administrative economic policies for landscapes, population settlement, and social control'.[3] In the early twentieth century nascent ecological and organicist thinking informed plans to regulate the relations between the city and the country in order to manage the life and welfare of the proletarian subject. The pathologies of the industrial city – its corruptive impact upon the moral and physical health of the worker, and its consequently negative influence upon productivity – were to be remediated through reintegration with the natural environment. This enterprise was to be facilitated through the design of environs affording access to sunlight and fresh air, to the ameliorative benefits of landscaped parks and gardens. Ebenezer Howard's Garden City is the prototype of this endeavour. It plots out relations between settlements and landscape. It coordinates social functions and prescribes their proper location within its territory. The Garden City, for Howard, is an essentially paternalistic project. In subsequent iterations its biopolitical powers are directed toward other agendas. It informs the anti-urban orientation of socialist planning in Soviet Russia and the machine-age modernization of the city envisaged by Le Corbusier. In 1920s Germany, the Garden City-inspired modernist *Seidlungen* of Ernst May and Bruno Taut seek to regulate the metabolic relations between the life of the subject and that of the soil. With the involvement of Leberecht Migge – himself inspired by the botanist Raoul Heinrich Francé's writings on soil fertility and organic farming methods – the residents of the settlements are provided with vegetable allotments from which to sustain themselves, and with instruction manuals on productively recycling their own bodily waste back into the same grounds.[4] If, in the interwar years the alienation of bodily waste from metabolic systems is abjured, then following the Second World War it is the psychological alienation produced by the city that is seen to require organic remediation. In his *Organische Stadtbaukunst: Von der Grosstadt zur stadtlandschaft* (Organic Urban Design: From the Metropolis to the City Landscape) of 1948, the German urban planner Hans-Bernhard Reichow argues that the urban subject is mentally and spiritually estranged by the chaotic and overcrowded conditions of the metropolis.[5] As a corrective, the space of the city should be opened up to form a landscape, a *stadtlandschaft* in which the subject can properly apprehend its environment as an organic whole of which it is an integral part.

Operators' manuals

The environmental consciousness that develops over the course of the 1960s makes the perspective of such projects appear parochial by comparison. Organic, metabolic and ecological processes are now conceived as operating at a planetary scale. The scope of the environment expands from the regional to the global, from Patrick Geddes's 'valley section' to Apollo 8's 'Earthrise'. The emergent global consciousness is driven, in part, by the space-time compression effects of advances in telecommunication and transportation, but also by an increasing awareness of the interrelatedness of ecological systems and their susceptibility to human intervention. The Earth is both smaller and more fragile than had previously been understood. On the cusp of this apparent crisis certain figures – Buckminster Fuller, John McHale, Ian McHarg, Reyner Banham – essay responses that will come to exert a significant influence on landscape, architectural and urban design. They promulgate ecological, organic, environmental and other models supposed to guide the conduct of the designer in addressing this crisis.

The crisis is also seized as an opportunity. The turn to the environment facilitates a turn against modernism in design. Modern architecture and urban planning are denounced as ill-informed projects of human hubris, unsympathetic endeavours in the manipulation of nature toward coldly rational ends. Prometheus is turned from hero to villain. Modes of mastery are to be surrendered to the laws of general systems theory. The creative agency of the designer is downgraded as the locus of design passes from subject to environment. At the same time, new allegiances are sought out. The roles earlier established for architecture, landscape and urban design within and for the nation state are made to appear anachronistic, as are the politics in which these were implicated. Designers, now recast as steersmen and cybernetic pilots, might do better to ally themselves with a universal humanity, to 'man' and his attunement with a correspondingly universalized nature. This prospect is explored, especially, in the allegiances forged between architecture and the counterculture in the 1960s and 1970s. '"Man"', as Douglas Murphy notes in his *Last Futures*, 'becomes a word that universalises humanity through a sense of cosmic distance, where divisions indistinguishable and patterns of behaviour common to all . . . "man" takes on a conspicuousness for its stressing of a (not unproblematic) universality, a world beyond struggle and conflict.'[6]

For Buckminster Fuller, man's best hope lay in forecasting the direction in which a global humanity was evolving and steering it, accordingly, with the guidance of general systems theory. In his *Operating Manual for Spaceship Earth* of 1969, he argues that nature itself has designed this spaceship and programmed its passengers for the mission evolutionary progress has long ago mapped out for them: 'What nature needed man to be was adaptive in many if not any direction; wherefore she gave man a mind as well as a co-ordinating switchboard brain.'[7] Spaceship Earth, in Fuller's account – sounding not unlike some version of 'intelligent design' – was 'invented' and 'superbly designed'.[8] This spaceship, carrying and nurturing us on its pre-plotted course, is also strategically programmed. The historical point at which 'man' becomes conscious of the crisis to which he is subjecting it, is precisely – as if by predestination – that at which Spaceship Earth reveals to him its true nature so as to enable its passengers to continue safely on their journey: 'It is therefore paradoxical but strategically explicable . . . that up to now we have been mis-using, abusing, and polluting this extraordinary chemical energy-interchanging system for successfully regenerating all life aboard our planetary spaceship.'[9] Computation comes to our rescue: 'A new, physically uncompromised, metaphysical initiative of unbiased integrity could unify the world. It could and probably will be provided by the utterly impersonal problem solutions of the computers.'[10] Guided by superior processing powers humans can let go of the wheel and enjoy the journey. The old ideologies that divided humanity are rendered redundant in the face of an unbiased and universal computation that will solve our crises for us, and in accord with the *telos* Spaceship Earth has been designed to realize. Capitalism and socialism are 'extinct'.[11] The turn to computation, the technological fix, is not against nature because nature is understood by Fuller as programmed, as an essentially cybernetic and universal system. The human brain is a 'co-ordinating switchboard' to be steered toward acceptance of the 'generalized principles governing the universe' by which it too operates.[12]

John McHale's *The Ecological Context* of 1971 likewise identifies the point at which the ecological crisis first appears as simultaneously the point at which its solutions are made manifest: 'Now capable of destroying his own species many times over through consciously contrived means of war, it suddenly becomes apparent that his [man's] largely unconscious, and uncontrolled, exploitation of the earth's resources may similarly render the planet unliveable.'[13] Humanity is poised at an epochal point of transition. Its emerging consciousness of global-scale ecological interconnectedness offers the key to its successful passage through this critical moment. The 'conceptual and value shift – from the local study of plants, wildlife and their surroundings to one which suggests responsibility for the planet as life-space – is in accord with the kinds of changes in human consciousness and conceptuality that are already underway.'[14] These changes in consciousness point to an awareness of the universal principles of ecology; a universality that captures the 'social behaviours of man' as much as the biological behaviour of plant and animal species. In light of this realization McHale, like Fuller, declares existing political paradigms – in fact politics as such – outmoded. The turn to ecology is the 'real' revolution, though the term preferred is 'evolution'. McHale also follows Fuller in his enthusiasms for cybernetics and the 'steering' of complex systems. At the planetary scale of 'large-scale scale systems with many and variable factors' man and nature cannot and should not be coerced.[15] Through the timely arrival of computational technologies, a world now understood according to the correct concepts will be cybernetically steered 'in a more positive, efficient, and naturally advantageous manner'.[16]

The landscape architect Ian McHarg similarly sought out a universal model through which his discipline might address the apparent crises of this period. His 1966 essay 'Ecological Determinism' opens with an account of the huge increases in population and urbanization then predicted for the United States. Attempts to plan for this development through the social and economic sciences, argues McHarg, only succeed in placing the city on the path to becoming a 'necropolis' of social and environmental breakdown.[17] Only through the recognition of 'the implications of natural process upon the location and form of development' can this fate be averted.[18] 'We need a general theory which encompasses physical, cultural and biological evolution', he writes in his 1968 essay 'Values, Process and Form'.[19] This need is answered by the science of ecology, which should now constitute the basis from which the landscape architect engages with the environment. 'The role of man', he writes, 'is to understand nature, which is to say man, and to intervene to enhance its creative process'.[20] The locus of creative agency, as with Fuller and McHale, is transferred from 'man' to nature. 'Man', likewise, is understood as adequately accounted for within the universal model of ecology. Design must work *with*, rather than *on*, nature – as in the title of McHarg's influential publication of 1969, *Design with Nature* – so as to enhance its innately creative evolution.[21]

In the field of architecture, Reyner Banham proposed that the discipline focus its concerns on the environment and its management. Influenced by Fuller, he argued, in his *The Architecture of the Well-Tempered Environment* of 1969, for 'the close dialogue of technology and architecture' toward this end.[22] Banham's affinities with countercultural sensibilities, and with the emancipatory promise of an environmentally attuned architecture, are especially apparent here. The profession has been tied, traditionally, to monied interests and state patronage. These ties have taught it to design 'enclosed spaces framed by massive structures'.[23] 'Civilized' architecture, claims Banham, is unable to conceive of 'free' or 'unlimited' space.[24] In other, non-Western traditions, such as those of nomadic peoples, he argues, habitation was managed without resorting to the structural segregation of humans from their immediate environment. Contemporary architecture, deploying the latest technologies, might also now cast off the dead weight of its civilizational orthodoxy and provide for the enjoyment of similar environmental liberties. Banham's enthusiasms for the frontier spirit of the campfire speak of his proximity to the libertarian values of the counterculture, especially as espoused by

Stewart Brand and as promoted in the latter's *Whole Earth Catalogue*. The technologically facilitated liberty to be at one with the immediate environment promoted by Banham is captured, famously, in the photograph of him riding his Bickerton bicycle across a Californian salt flat taken by Tim Street-Porter in 1981. The architectural historian wrote of the rapturous nature of this experience in his *Scenes from American Deserta*: 'Swooping and sprinting like a skater over the surface of Silurian Lake, I came as near as ever to a whole-body experience equivalent to the visual intoxication of sheer space that one enjoys in America Deserta.'[25] In his *Los Angeles: The Architecture of the Four Ecologies* of 1971 Banham glossed the experience of environmental immersion offered by the Los Angeles freeway system in similarly transcendent terms. The cybernetic traffic system operates at a level of complexity that demands the 'willing acquiescence' of the driver to its systemic commands.[26] The human is too slow to calculate and respond effectively in this environment and must give up its own agency to that of the system's superior processing power. In recompense, however, the Los Angeleno is delivered over to the 'mystical' experience of the freeway, released in relinquishing control to the opportunity to be 'integrally identified' with the urban landscape.[27]

Environmental Disneyland

The writings of Fuller, McHale, McHarg and Banham are representative of a broader current of enthusiasm for all things environmental, ecological and cybernetic in design during this period. This turn was not, however, unopposed. At the Design and Environment Conference, organized by Banham and held at Aspen in 1970, the radical French architectural group Utopie delivered a statement in objection to the emerging doxa of 'environments' and 'environmentalism'. Written by Jean Baudrillard and delivered by a delegation calling itself 'The French Group', their paper was titled 'The Environmental Witch-Hunt'. In it was written:

> The burning question of Design and Environment has neither suddenly fallen from the heavens nor spontaneously risen from the collective consciousness: It has its own history. Professor Banham has clearly shown the moral and technical limits and the illusions of Design and Environment practice. He didn't approach the social and political definition of this practice. It is not by accident that all the Western governments have now launched (in France in particular for the last six months) this new crusade, and try to mobilize people's conscience by shouting apocalypse.[28]

Baudrillard and the Utopie group understood the conference as representative of an emerging 'environmental ideology', a smokescreen masking a deeper crisis within capitalism. 'Aspen', wrote Baudrillard, 'is the Disneyland of environment and design' – a 'Utopia produced by a capitalist system that assumes the appearance of a second nature in order to survive and perpetuate itself under the pretext of nature'.[29] Baudrillard insists on understanding the turn to the environment in relation to the escalation of the class struggle, as represented, most visibly and emblematically, by *les événements* of May 1968, but also encompassing a broader wave of protests, uprisings and strikes elsewhere in Western Europe and the US at this time. This period has also come to be understood, retrospectively, as marking a crisis point within the existing Fordist model of capitalism, a crisis of accumulation that engendered the shift toward post-Fordist and neoliberal modes of production and accumulation. Yet the turn to the environment, and with it to ecology and cybernetics, is not only, and not most significantly, ideological. The environment is not so much a means of distraction as it is a new mode of control that will serve to resolve the crisis for capitalism.

The potential for the exercise of power through environmental means was early recognized by Marshall McLuhan. Typically misread as a cheerleader for the cybernetic enhancement of human capacities, McLuhan was unambiguous in his concerns over the impact of the emerging cybernetic environment – one of mass and globally extended communication systems – on the human subject.[30] In an essay of 1967 published in the architectural journal *Perspecta*, 'The Invisible Environment: The Future of an Erosion', McLuhan suggests that all environments, as such, possess the capacity to function as highly effective modes of propaganda. Their 'action . . . is total and invisible, and invincible'.[31] The 'environment' is not, he writes, necessarily 'bad', but its 'operation upon us' is always 'total and ruthless'.[32]

The most prescient analysis of the ways in which power was turning to environmental means to effect new modes of control is given by Michel Foucault. In the lectures he presented at the Collège de France in 1978–1979, gathered under the title *The Birth of Biopolitics*, Foucault frequently employs the term 'environment' to articulate the shift in the mechanisms of power that first appears in the aftermath of the Second World War.[33] In place of disciplinary power, and in reaction to its most extreme and totalitarian expressions, there appears on the horizon:

> [T]he image, idea, or theme program of a society in which there is an optimization of systems of difference, in which the field is left open to fluctuating processes, in which minority individuals and practices are tolerated, in which action is brought to bear on the rules of the game rather than on the players, and finally in which there is an environmental type of intervention instead of the internal subjugation of individuals.[34]

This image of environmental intervention is formulated from within the economic philosophy of German ordo-liberalism and Austro-American neoliberalism. It is premised on the conception of the human subject as a fundamentally entrepreneurial being, a *homo economicus*. Rather than disciplined to behave in accord with the dictates of the authority of the state, this subject must be at liberty to pursue its entrepreneurial interests within an unbounded environment of opportunities for exchange, competition and accumulation. It might appear that the term 'environment', here used in reference to the milieu of the market, has little real relation to the sense in which it is used within design. However, neoliberal thought models the market, and the action of its subjects, according to precisely the same cybernetic models as does design at this time. Though not addressed by Foucault, the writings of the economist Friedrich Hayek – the key intellectual within what has been referred to as the 'neo-liberal thought collective' – reveal the influence of cybernetics and systems theory in the development of neoliberalism.[35] Cybernetics serves, for Hayek, to underwrite his contention that individuals are liberated to act in accord with their true entrepreneurial natures only by being immersed in larger environments.[36] These individuals, however, cannot hope to exercise rational control of these environments. This desire leads only to totalitarianism. It must be renounced in favour of acknowledging, and giving oneself over to, the superior organizational and processing power of the system that spontaneously orders all the elements within the environment. This spontaneous and self-generating environmental order – of which the economic market is exemplary – operates as an immanent mode of governance that appears to transcend the political and serves to guide us 'by habit rather than reflection'.[37] Governmental powers are now invested in the mutually legitimating discourses of the economic and the environmental. Both play out their hands invisibly.

Neoliberalism, design and the artifice of the immediate

The development of neoliberal thought and environmental design coincide historically and conceptually. Though not identical, they occupy common ground. Human agency is delegitimated, its powers reallocated to systemic orders – ecology, environment, nature, the way of the world. At the planetary scale requisite to conceiving of ecological systems, and to understanding the operations of the globalized market, matters are too complex to be directed by human creativity or intelligence. We are, in our limited human capacities, claims Hayek, 'necessarily ignorant'.[38] Neoliberalism and environmental design are, consequently, in agreement on the dangers of human hubris. Natural and computational systems, of which 'man' is now an integral but in no way privileged component, are better able to calculate and regulate our relationship to the environment. Understood to operate according to universal principles, they are presented as the true agents of design and creativity. Their processing powers alone can handle the now globally scaled degrees of complexity, from the bottom up.

There are common grounds, shared affirmations of greater realities to which we are supposed to have immediate access. Realities we are called upon, in light of their universality, to surrender our narrow political allegiances and renounce self-interest. These commonalities are sufficient to have design function biopolitically, to have it operate as an instrument of neoliberal power. Though not immediately acted upon, this potential has begun to be realized of late. Ascendant neoliberalism, in its projects for remaking subjectivity in its own image, has found employment for designers. Networked spaces of labour, learning landscapes, malls without walls, eco-cities, smart cities. All designed as open environments, interactive zones for the practice of behavioural flexibility. Design has become especially amenable to these purposes following its return to cybernetics and systems theory after the postmodern interregnum of the late 1970s and 1980s. Since the early 1990s the field has become increasingly attracted to universalizing theories and their capacity to underwrite its practices. Not only cybernetics, but theories of complexity, self-organization and emergence have come to occupy a central place in its discourse, one that works to legitimate its claims to be now in touch with the systemic, ecological and environmental realities of the world. At the same time, the turn toward these realities has been used as the basis for a disavowal of the political and the critical. With unmediated access to the way of world, and one whose very survival is at stake, these appear as outmoded concerns.

Landscape urbanism has been broadly allied with and contributed to this discourse. In his influential essay 'Field Conditions', Stan Allen – a figure significant not only to the development of Landscape Urbanism but notable for his contributions to the 'post-critical' discourse in architecture – locates the emergence of what he identifies as a generalized shift from 'object to field' amidst the science, technology and culture of the post-war period of the twentieth century.[39] '[F]locks, schools, swarms and crowds' are described as 'field phenomena'. Their behaviour can be understood through the science of 'chaos theory'.[40] The scientific understanding of these phenomena places design 'in contact with the real'.[41] In an essay titled 'Fuzzy Thinking', Andrea Branzi, another figure of significant influence within landscape urbanism, argues that the complexity of nature revealed by recent developments in mathematics presents an 'evolved model to imitate in the process of building the new', one that constitutes a 'new naturalism'.[42] In this scenario the critical agency of the designer is surrendered to that of the natural phenomena whose ways he or she is merely required to mimic, whose immanent order serves as an operational imperative.

If landscape-related design, and design in general, is to recover the kind of critical agency relinquished through its engagement with environmental and ecological models, it might begin by recognizing the place of artifice within its discourse and practice; 'artifice' in the sense of an art of making, one that involves skill, even cunning, in the crafting of devices, and that is understood as a peculiarly human practice. Artifice is not an environmental given or a universal principle immanent to all nature. In fact, the very models through which the world is understood to be universally complex, self-organizing and spontaneously ordered are themselves forms of artifice. Even, and especially, the notion that we can bypass all forms of mediation and gain immediate access to the reality of these operations is artificial. Though not necessarily false or duplicitous, these are necessarily mediated – culturally, historically, economically, linguistically, socially – ways of conceiving and representing the world. Recognizing the forms of artifice implicated in environmental and ecological models doesn't require that these are simply cast aside, but that the kind of work that they might do, the kinds of skill and cunning implicated in their production, are reflected upon in light of the purposes and interests they might serve. Designers, recognizing the place of artifice within their own practice, might be more ready to acknowledge and reflect upon the significance of their agency. Without the unquestioned authority of universal models of reality to hide behind, and disabusing themselves of the myth of universal 'man', they might have cause to reflect and act upon the politics of their practice.

Notes

1 Waldheim (2016, ii, 2, 4, 6, 14, 50, 55, 93, 140, 171).
2 Foucault (2004, 253).
3 Ankar (2001, 2).
4 For a thorough historical account of Migge's involvement in the design of the German modernist *seidlung*, see Haney (2010). See too, for an analysis of the relationship between modernism and the organic, environmental and biological, Botar and Wünsche (2011).
5 Reichow (1948).
6 Murphy (2015, 15). See too chapter 5, 'Cybernetic Dreams', pp. 105–136, on the relationship between design and the counterculture.
7 Fuller (2008/2011, 25).
8 Ibid., 59.
9 Ibid., 59–60.
10 Ibid., 45.
11 Ibid., 48.
12 Ibid., 66.
13 McHale (1971, 1).
14 Ibid., 2.
15 Ibid., 89.
16 Ibid., 89.
17 McHarg and Steiner (1998, 40).
18 Ibid., 40.
19 McHarg (1998, 63).
20 Ibid., 59.
21 McHarg (1995).
22 Banham (1984, 28).
23 Ibid., 21.
24 Ibid., 20.
25 Banham (1982, 99).
26 Banham (2001, 199).
27 Ibid., 197.
28 Baudrillard, J./The French Group (1974, 208).
29 Ibid., 210.
30 McLuhan (1967, 163–167).
31 Ibid., 164.
32 Ibid., 164.
33 Foucault (2008). For an incisive reading of Foucault's analysis of environmental modes of control in relation to architecture and design, see Shvartzberg (2015, 181–206).
34 Ibid., 259–260.
35 Mirowski (2013).
36 For a more extensive analysis of the common ground shared between neoliberalism and design thinking in relation to systems theory and cybernetics, see Spencer (2016).
37 Hayek (2006, 66).
38 Hayek, in *The Constitution of Liberty*, states that the primary requisite for understanding society is that 'we become aware of men's necessary ignorance of much that helps him to achieve his aims' (p. 21).
39 Allen (1999).
40 Ibid., 99.
41 Ibid., 92.
42 Branzi (2006, 29).

References

Allen, S. (1999) 'Field Conditions', in *Points + Lines: Diagrams and Projects for the City*. New York: Princeton Architectural Press.
Ankar, P. (2001) *Imperial Ecology: Environmental Order in the British Empire, 1895–1945*. Cambridge, MA: Harvard University Press.
Banham, R. (1982) *Scenes in America Deserta*. London: Thames & Hudson.
Banham, R. (1984) *The Architecture of the Well-Tempered Environment*, 2nd edn. Chicago, IL: University of Chicago Press.
Banham, R. (2001) *Los Angeles: The Architecture of the Four Ecologies*. Berkeley and Los Angeles, CA: University of California Press.
Baudrillard, J./The French Group (1974) 'The Environmental Witch-Hunt', in R. Banham (ed.) *The Aspen Papers: Twenty Years of Design Theory from the International Design Conference in Aspen*. New York: Praeger.
Botar, O.A.I. and Wünsche, I. (eds) (2011) *Biocentrism and Modernism*. Farnham, Surrey and Burlington, VT: Ashgate.
Branzi, A. (2006) 'Fuzzy Thinking', in *Weak and Diffuse Modernity: The World of Projects at the Beginning of the 21st Century*, trans. Alta Price. Milan: Skira.
Foucault, M. (2004) *Society Must Be Defended: Lectures at the Collège de France*, trans. David Macey. London: Penguin.
Foucault, M. (2008) *The Birth of Biopolitics: Lectures at the Collège de France, 1978–79*, trans. M. Senellart. Basingstoke, UK and New York: Palgrave Macmillan.
Fuller, R.B. (2008/2011) *Operating Manual for Spaceship Earth*. Baden, Switzerland: Lars Müller.
Haney, D. (2010) *When Modern Was Green: Life and Work of Landscape Architect Leberecht Migge*. London and New York: Routledge.
Hayek, F.A. (2006) *The Constitution of Liberty*. Abingdon and New York: Routledge.
McHale, J. (1971) *The Ecological Context*. London: Studio Vista.
McHarg, I.L. (1995) *Design with Nature*. Steelville, MO: San Val, Incorporated.
McHarg, I.L. (1998) 'Values, Process and Form', in I.L. McHarg and F.R. Steiner (eds) *To Heal the Earth: Selected Writings of Ian L. McHarg*. Washington, DC and Covelo, CA: Island Press.
McHarg, I.L. and Steiner, F.R. (eds) (1998) *To Heal the Earth: Selected Writings of Ian L. McHarg*. Washington, DC and Covelo, CA: Island Press.
McLuhan, M. (1967) 'The Invisible Environment: The Future of an Erosion', *Perspecta* 11, pp. 163–167.
Mirowski, P. (2013) *Never Let a Serious Crisis Go to Waste: How Neoliberalism Survived the Financial Meltdown*. London: Verso.
Murphy, D. (2015) *Last Futures: Nature, Technology and the End of Architecture*. London and New York: Verso.
Reichow, H. (1948) *Organische Stadtbaukunst: Von der Grosstadt zur stadtlandschaft*. Braunschweig: Georg Westermann Verlag.
Shvartzberg, M. (2015) 'Foucault's "Environmental" Power: Architecture and Neoliberal Subjectivization', in Peggy Deamer (ed.) *The Architect as Worker: Immaterial Labour, the Creative Class, and the Politics of Design*. London: Bloomsbury, pp. 181–206.
Spencer, D. (2016) *The Architecture of Neoliberalism: How Architecture Became an Instrument of Control and Compliance*. London: Bloomsbury.
Waldheim, C. (2016) *Landscape as Urbanism*. Princeton, NJ and Oxford: Princeton University Press.

AFTERWORD

Landscape's agency

Don Mitchell

'A working country,' Raymond Williams asserted in *The Country and the City*, 'is hardly ever a landscape.' When 'landscape' is understood in its artistic (and often its design) sense, Williams is absolutely right: landscape necessarily 'implies separation and observation', as he argues.[1] More than that, though, landscape is not just separate from work; it actively obfuscates work. If, after the Fall, humans suffered the 'curse' of having to earn their bread in the sweat of the brow (as the Bible has it), then the creation of the English countryside landscape was accomplished, in art and then seemingly in the landscape itself, by a:

> magical extraction of the curse of labour [which] is in fact achieved by the simple extraction of the existence of labourers. The actual men and women who rear the animals and drive them to the house and kill them and prepare them for meat; who trap the pheasants and partridges and catch the fish; who plant and manure and prune and harvest fruit trees; these are not present; their work is all done for them by a natural order.[2]

With first the industrial, and now the urban revolution, this 'magical extraction' has hardly disappeared.[3] Our meat, fruit and vegetables continue to magically appear in our stores and restaurants; electricity magically appears in our homes that magically appeared on our realtors' websites and then magically before us as our feet take us across the threshold (though after our mortgages have been secured); our fallen leaves are magically blown away and our grass is magically cut; roads and lanes magically appear before the rolling wheels of our cars; petrol magically appears at the pumps (even if we know that war is necessary at its origins we pay no mind to what happens once the crude is in the pipelines); waste is magically whisked away in the middle of the night and then made to disappear altogether; T-shirted tech-wizards stand on cavernous stages before adoring audiences pretending to magically conjure out of thin air the wonderful devices you are about to hold in your hand and that now let us magically see any landscape we might want.[4]

It takes no thought at all to know this is all an illusion. Landscape is *nothing but* work (as Raymond Williams well knew). It's a *produced* space. Landscape's genius, landscape's own agency, is that it is forever masking the work that makes it. That's the very point of many designed landscapes, like many of those examined in this book. But it is just as much simply what all kinds of landscapes do – from landscapes of suburban tract homes to landscapes of deindustrialization (which is never *only* abandonment; always *especially* a product of specialized labour) and from gentrified inner-city landscapes to awe-inspiring natural landscapes preserved in wilderness parks (nature never goes away, but natural parks are hardly natural). It takes no thought, but most of the time we are quite content with the illusion. And why not: it makes life bearable (even as landscapes are sometimes wrenched out of completely unbearable lives – or are made only by making life unbearable by those who are displaced so our comfort can be vouchsafed). Aesthetically, socially, politically, we – we for whom landscapes are made – revel in landscape's agency, its genius, its magic: the order it imposes on space and its ability to hide from us the conditions of its own production. This magic is deeply consequential: it sets the stage for any further labour, any further social practice of any kind. Landscape is a machine for alienation.

Landscape's magic – its agency, its alienating force, its fetishizing power – can be *revealed* in lots of ways. Any number of radical art and architecture historians, literary scholars and anthropologists, social historians and geographers of a Marxist stripe have set themselves the task of defetishizing the landscape, revealing the work and processes that make it. From the pioneering work of John Barrell, Ann Birmingham and (a bit more recently) Elizabeth Helsinger in art history, to Raymond Williams and WJT Mitchell in fields all their own, and on to geographers stretching from Denis Cosgrove and Stephen Daniels through to Tom Mels and Clayton Rosati, the goal of much landscape research has been to pull back its mystifying veil and expose the processes of its inner production – but then also to show how and why that veil matters so deeply to what landscape *is*, what it does, what it means (socially and ideologically), and therefore how it continues to exert social force. In so doing, such revelatory work also seeks to uncover and make plain for all to see not only the relations of labour that go into landscapes' making but especially the labourers themselves who are, in essence, the agents of landscapes' own agency.[5]

But if landscape's fetishizing agency can be revealed in many ways, it can be *dispelled* in only one way: by taking it and making it something else altogether; by pulling it up by its roots and giving it a good shake; by transforming it from a landscape to a *public space*. The argument I am making here is based on a particular way of understanding public space. Public space is not necessarily space owned by the state, and it is not necessarily even space specifically designed to be open and accessible. Rather it is space made public through the practices and political actions of people using it, often in opposition to its intended purpose. Henri Lefebvre describes the distinction I am making as a dialectic between abstract space and differentiated space.⁶ Abstract space is space made commensurable – exchangeable. Its purpose is to support the dominant or hegemonic social relations of the time, which in our time are capitalist and thus demand the exploitation, alienation and obfuscation of labour. Against the constant production of abstract space (through the circulation of capital in the built environment and David Harvey has done so much to theorize),⁷ struggles erupt to differentiate it. Lefebvre argues that only the class struggle (which he defines with appropriate capaciousness) prevents abstract space from 'papering over' the whole world, and he is correct, at least if we understand this struggle to be multifaceted and aimed at systems of political and social dominance not just systems exploitation within the workplace.⁸ These fights for differentiation, *in the moment of struggle*, turn landscape – an abstraction of space – into public space, upsetting settled practices and relations of power and making them into something else. Eventually the space is then either *reclaimed* or *reformed*.

These processes can be glimpsed nearly anywhere, but Union Square in New York City might be a paradigmatic example. Union Square has, of course, long been a central gathering place in the city.⁹ After Lincoln declared war against the South, it hosted the largest rally to date in America. It was the site of New York's first Labor Day rally in 1882. And it has always been a popular gathering place for May Day demonstrations, street speakers and agitators of all types. But during World War II, street speakers were banned and the park began a long decline. In the 1980s, the square was redeveloped as part of the city's effort to assure parks would become an asset to real estate. Parks were to be part of a new city of commensurability, a landscape in which exchange value trumped use value and any place failing to contribute to an overall rising rent surface had to be reclaimed and remade. Landscape's agency was to be pressed into the service of profit-making. A new, private Union Square Partnership was created to oversee redevelopment. But, according to the great urbanist Marshall Berman, the reconstructed square – a landscape par excellence, designed precisely to efface work and difference and to hide the facts of its own social relations – did not 'catch on as a public space' until September 11, 2001:

> Then, abruptly, it was flooded with candles, flowers, missing person signs, poems and drawings. Some art students unrolled a scroll of paper three feet wide and several hundred feet long. A great assembly of people gathered round the scroll, and wrote radically contradictory messages and meditations. Overnight, Union Square became the city's most exciting public space: a small-town Fourth of July party combined with a 1970s be-in.¹⁰

In the first days after the attack, Union Square – the gathering area closet to the Lower Manhattan 'exclusion zone' from which members of the general public were banned – became a vital site of public mourning, as well as a place to post photos of missing loved ones. Soon a new social geography developed in the park, with street kids and homeless people moving into the centre; activists and musicians taking over the southern steps; and Christians gathering over on the east side.

People descended on the park in their thousands, with many staying the night, replacing and relighting candles. Decorative columns were turned into shrines; George Washington and his horse were bedecked with flowers, signs and peace symbols. Daily parades behind American flags were held. Floral representations of the Twin Towers bloomed in the flower beds. By the weekend, 'the park had become an impromptu outdoor festival', as the *New York Times* architecture critic put it. Landscape's agency had been interrupted: it was now functioning 'as urban designs are [often] conceived to do: bring strangers together on common ground, people who otherwise might never have met, people who would not have bothered to notice one another on the street'.[11] All of a sudden the *work* that makes and remakes a space public was plain for all to see – and there was nothing magical about it (as magical as Union Square came to be for those who spent time in it).

Within a week Union Square had become a place for debate and proselytizing, mourning and organizing. And as the month wore on, it became an even more important place for those who, as the George W. Bush administration rushed towards war, demanded peace. A peace march was planned for Friday, September 21, to begin at the square and head north to the Armed Forces recruiting station in Times Square. To the city, this was intolerable. Early morning the day before the march, the city moved into the square. Police and city employees pushed out the homeless and street kids; other crews dismantled the posters, photographs and paintings that adorned so many surfaces. Yet other workers power-sprayed the candle wax and scrubbed Washington and his horse clean. When all this was done, a fence was erected all the way around the square, closing off its heart. New York's most vibrant public space – a space made newly public as thousands of people expressed grief, outrage, hope for the future, and more importantly organized to assure a different kind of world might come into being – was destroyed in a single morning. Through the work of reclamation, order was restored. Abstract space was protected from its differentiation. Union Square, like Lower Manhattan, was quickly returned to its proper function within the capitalist city.

The point is this: even if only temporary, such *taking* of space and differentiating it interrupts landscape and, by making it a *public* space, dispels its fetishizing agency. Something else has to arise in its place, even if that else is a concerted restoration of what was there before. After a fashion, I think this is just the point Ed Wall is making in his chapter in this volume, only he enlarges it asking if we have in fact achieved a 'post-landscape' condition. And given the reconquest or restoration of landscape that always remains possible, he asks the right next question: *how* is a post-landscape condition – one where landscape's traditional agency is dispelled – to be sustained? But I think there is yet a further question that also needs to be asked: And with what should we replace it? For, if not fetishization, some sort of wilful blindness to the raw, tumultuous, ever-in-flux, never concretized, now over-exposed social relations that actually make space is, in fact, necessary. None of us can live a totally exposed, totally indeterminate life. We need order. We need fixity. We need a degree of stability. We need work to solidify into something – to become internalized in something, just as now happens with commodities or landscapes.

So we need to talk (a lot) about what these post-commodities and post-landscapes might look like (literally). And, particularly, we need to start figuring out what productions of space can both dispel landscape's fetishizing agency and create ordered, reasonably fixed, open, places that refuse the exploitations and oppressions that make landscape possible. How might we turn landscape's agency into something else altogether? The chapters in this book take an important step in opening up this conversation. But those who gathered at Union Square after 9/11 – or those at Occupy London that Wall writes about – are doing even more: they just might very well be showing us the way: the way not toward some 'fleeting ephemerality' of landscape (as Wall puts it) but rather toward a new kind of post-landscape order, one being *worked* out on the ground.

Notes

1 Raymond Williams (1973) *The Country and the City*. Oxford: Oxford University Press, p. 120.
2 Ibid., 32.
3 Henri Lefebvre (2003) *The Urban Revolution*. Minneapolis, MN: University of Minnesota Press.
4 There is a whole cottage industry in geography staffed by theorists dedicated to the proposition that the landscape is magical and immanent. They call themselves 'new materialists' or 'new phenomenologists' and they vibrate wildly with excitement as the landscape apparently only comes into being through their immediate experience of it. As examples, see J.W. Wylie (2005) 'A Single Day's Walking: Narrating Self and Landscape on the Southwest Coast Path', *Transactions of the Institute of British Geographers* 30, pp. 234–237; Mitch Rose (2002) 'Landscapes and Labyrinths', *Geoforum* 33, pp. 455–467; David Bissell (2010) 'Vibrating Materialities: Mobility-Technology-Body Relations', *Area* 42, pp. 479–486.
5 For reviews, see Don Mitchell (2003) 'California Living, California Dying: Dead Labor and the Political Economy of Landscape', in K. Anderson, S. Pile and N. Thrift (eds) *Handbook of Cultural Geography*. London: SAGE, pp. 233–248; Don Mitchell (2005) 'Landscape', in D. Sibley, P. Jackson, D. Atkinson and N. Washbourne (eds) *Cultural Geography: A Critical Dictionary of Key Ideas*. London: IB Taurus, pp. 48–56; and Don Mitchell and Carrie Breitbach (2011) 'Landscape, Part 1', in J. Agnew and J. Duncan (eds) *The Wiley-Blackwell Companion to Human Geography*. Oxford: Blackwell, pp. 209–220.
6 Henri Lefebvre (1991) *The Production of Space*. Oxford: Blackwell.
7 David Harvey (1982) *The Limits to Capital*. Chicago, IL: University of Chicago Press; David Harvey (1989) *The Urban Experience*. Oxford: Blackwell.
8 Lefebvre (1991, 55).
9 The analysis here draws on one I make in a vignette on Union Square in Neil Smith and Don Mitchell (general editors) (forthcoming 2018) *Revolting New York: How 400 Years of Riot, Rebellion, Uprising, and Revolution Make a City*. Athens, GA: University of Georgia Press.
10 Marshall Berman (2002) 'When Bad Things Happen to Good People', in Michael Sorkin and Sharon Zukin (eds) *After the World Trade Center: Rethinking New York City*. New York: Routledge, pp. 1–21, quotation from p. 11.
11 Michael Kimmelman (2001) 'In the Square, a Sense of Unity: A Homegrown Memorial Brings Strangers Together', *New York Times*, 19 September.

INDEX

abstract space 190
Actor-Network-Theory (ANT; Latour) 19
Adams, Ross Exo 4
Additional Protocol to the Convention (American) 60, 63
Agamben, Giorgio 79, 80
agriculture, urban 131–43
Akerman, Chantal 16*ph*
Alexander, Christopher 107
Allen, Stan 185
Alphand, Jean-Charles 72
Ambridge, Pennsylvania 22–6, 23*ph*, 30
American Bridge Company 23–4, 26
American Convention 59–60
Anker, Peder 178
Arcadia 66
Architecture of Neoliberalism, The (Spencer) 3
Architecture of the Well-tempered Environment, The (Banham) 181
artifice 177–8, 185
Asian Coalition for Housing Rights (ACHR) 168
Awas Tingni v. Nicaragua 57–8

Baan Eua Arthorn Programme 168
Baan Mankong housing and community upgrading programme 164, 167–74, 169*fig*
Bang Bua Thong District *khlong* (canal) 166*fig*
Bangkok 164–76
Bangkok Noi Canal 167
Bangkok Transit Map 166*fig*
Banham, Reyner 179, 181–2
Barrell, John 151, 189

Baudrillard, Jean 182
Bay Lexicon (Wolff) 35, 40–2, 41*ph*, 42*ph*, 47–9
Bender, Barbara 144, 145, 146, 152, 156, 162
Bennett, Jane 8
Berger, John 105, 152, 159
Berman, Marshall 190
Berque, Augustin 9–10
Biemann, Ursula 15*ph*
biopolitics 70, 78–80, 85, 87, 178–9
bios 70, 79
Birmingham, Ann 189
Birth of Biopolitics, The (Foucault) 183
Black Sea Files (Biemann) 15
Bloomberg, Michael 29
Board of Commissioners of the Central Park (BCCP) 20
Bodwell Granite Company 20*ph*, 21–2, 22*ph*
Boonyabancha, S. 169
border zone 101–2
Botero, Giovanni 11
boundary versus border 101
Bourdieu, Pierre 120, 124
Brand, Stewart 81, 182
Branzi, Andrea 185
Brezhnev districts 92, 93
Buckley, John 25
Buckley v. the United Kingdom 55
Burns, Jim 105

Cabinet Magazine 161
California Delta 36, 42–6

Campkin, Ben 155
Canary Wharf 125
Cantillon, Richard 12
Central Park, New York City 19–22, 30
chaos theory 185
Chapman v. the United Kingdom 55
Chelas Valley 131–43
Cheung, Justin 38*ph*
Chin, Mel 155
chromated copper arsenate (CCA) 28
Churchill, Winston 122
Cities (Halprin) 109
city-as-a-forest 66–7, 74, 75
civil society 123–4
climate change 78–9
climate control 65, 67–70
closed cities 90–103; *see also* ZATOS
'CMWP' ('Classical Modern Western Paradigm') 10
Colbert, Jean-Baptiste 11
collages 159
commodity chains 19
'community architects' 170, 171*fig*
Community Organizations Development Institute (CODI) 164, 167–74, 169*fig*, 173*fig*
Corner, James 7, 8, 104–5, 150, 151, 152, 154, 155, 156, 159
Corps du génie 11
Cosgrove, Denis 1, 144, 150, 153, 155, 156–7, 189
Country and the City, The (Williams) 188
Cresswell, Tim 123, 165
Cronon, William 19
Cullen, Gordon 108
cultural heritage, landscape rights and 54–5, 62
Cultural Landscape Foundation 111
custom 123–4
cybernetics 180–1, 182, 183, 184–5

Daniels, Stephen 189
Dark Side of the Landscape, The (Barrell) 151
Décosterd, Jean-Gilles 68
De Landa, Manuel 102
Delta Primer 35–6, 42–6, 44*ph*, 45*ph*, 46*ph*, 47–9
Delta Protection Commission 46
democracy 117–30, 172
Denes, Agnes 78–9, 80, 81–2, 81*ph*, 83–4, 87
Depalle v. France 55
Design and Environment Conference 182
Design with Nature (McHarg) 181

differentiated space 190
Digestible Gulf Stream (installation) 68–9
Diller Scofidio + Renfro 29
Discovering the Vernacular Landscape (Jackson) 2
Downsview Park (Toronto) 66, 67, 72–5
doxa 120–2, 124
Dymaxion maps (Fuller) 83–4
Dymaxion World, The (Fuller) 85

'Earthrise' photograph 80–1, 179
École des ponts et chaussées 11, 12
Ecological Context, The (McHale) 180–1
'Ecological Determinism' (McHarg) 181
ecology of matter 8, 9
economic recession 146–7, 162
Ecuadorian Constitution 63
Eliot, Charles 48
'emergence' 177
EMF (Estudi Martí Franch) 157
'Energetic Geometry' projects (Fuller) 84
energy slave maps (Fuller) 78, 80, 83*ph*, 85–6
environmental consciousness 179–84
environmental justice 172, 174
'Environmental Witch-Hunt, The' (Baudrillard) 182
equality 118, 165, 171
Estado Novo (New State) 131, 133, 134
European Convention on Human Rights and Fundamental Freedoms 54
European Court of Human Rights 53, 54–7, 60–1, 62
European Landscape Convention (ELC) 52
événements, les 183
Exploratorium 40, 42

Fake Estates (Matta-Clark) 161, 161*fig*
Farrell, Terry 149
'Field Conditions' (Allen) 185
Field Operations 154
Fitzrovia 122
flash cards 40–2, 42*ph*, 43*ph*
flooding 34, 36
Flusser, Vilém 13
food justice movement 31
Forensic Architecture project 15–16
Forest Stewardship Council (FSC) 28, 29, 30
Foucault, Michel 12, 70, 79, 87, 183
Fox Islands, Maine 19–22, 30
frames/framing 153–4, 161
Francé, Raoul Heinrich 179
Fraser, Murray 4

Fraser, Nancy 156
'French Group, The' 182
Fresh Kills Park 2, 3, 154
Fuller, Buckminster 78–9, 80, 81, 82–7, 83*ph*, 179, 180, 182
'Fuzzy Thinking' (Branzi) 185

Gabinete Técnico de Habitação (GTH) 133, 134
G. and E. v. Norway 56
Garden City 178–9
Geddes, Patrick 179
Gell, Alfred 104–5
general interest of society 54–6, 57, 62
geo-materialist approach 19, 31
Goodge Street 122
Gottscho, Samuel H. 25*ph*
Gould, Stephen Jay 101
Gragg, Randy 111
Granite Ring 21
Greater New Orleans Urban Water Plan 39
Greek Society for the Protection of the Environment and Cultural Heritage 60
Gudeman, Stephen 119
Gutter to Gulf 35, 36–40, 37*ph*, 47–9

Habermas, Jürgen 156
Habitações de Renda Económica (Affordable Housing) 133
Halprin, Anna 107
Halprin, Lawrence 104–16, 106*ph*, 109*ph*, 110*ph*, 111*ph*, 113*ph*
Handolsdalen Sami Village v. Sweden 56
Haraway, Donna J. 86–7
Hardiejowski, Marc 38*ph*
Harmony Society 23, 26
Hartwick, Elaine 19, 31
Harvey, David 19, 146, 150, 152, 154, 155, 162, 190
Haussmann, Georges-Eugène 72, 74–5
Hayek, Friedrich 184
Heatherwick, Thomas 152
Helsinger, Elizabeth 189
Henry Hudson Parkway 24–5, 26
Herrick v. the United Kingdom 54
Hester, Randolph T. 105, 113
hexis 120–1
High Line, New York City 24, 26–30, 28*ph*, 29*ph*, 154
Highway Code 120, 125–7
Hingitaq 53 and others v. Denmark 56
Hirsch, Alison 104, 110
Hobbes, Thomas 11
Hoeferlin, Derek 37, 38

Hormonorium (installation) 68–9
hortas (allotment gardens) 135, 135*fig*, 136*fig*, 137, 138, 140
Howard, Ebenezer 178
human rights 52–64
Hunt, John Dixon 145
Hurricane Katrina 34, 36
Hutton, Jane 27*ph*

Imperial Ecology (Anker) 178
Incomplete Cartographies (Wall) 161, 159*ph*
indigenous land rights 56–9, 62
Ingold, Tim 120
Inter-American Commission on Human Rights 58
Inter-American Court of Human Rights 53, 56–60, 61–2
international law 52–64
'Invisible Environment, The' (McLuhan) 183
ipê lumber 26–30, 27*ph*
'Isometric Systems in Isotropic Space: Map Projections' (Denes) 81–2, 81*ph*, 84

Jackson, J.B. 2
Jacobs, Jane 91
Jade Eco Park 65, 66, 67–72, 67*ph*, 68*ph*, 69*ph*, 71*ph*, 75–6
James Corner Field Operations 29
John-Alder, Kathleen 109
Johnson, Peter 118
Johnson, Steve 102
justice: environmental 172, 174; food 31; social 165, 170; spatial 172

Kasson, John F. 118
Khrushchev districts 92, 93
Klee, Paul 105
Komara, Ann 104
Könkäma and 38 other Saami villages v. Sweden 56
Kozacioglu v. Turkey 54
Krasnoyarsk-26/Zheleznogorsk 92–3, 97–9, 102
Kyrtatos v. Greece 60, 61

Laban Notation (also Labanotation) 107
Landscape, Nature, and the Body Politic (Olwig) 1
Landscape as Urbanism (Waldheim) 177
landscape painting 144–5, 150, 151, 152
landscapes: as archive 8–9, 13–16; of Chelas valley 131–43; closed 90–103; as collective work 1–2; definitions

of 1; environmental recovery of 154–5; equitable 164–76; hybrid 34–51; international law and 52–64; neoliberalism and 177–87; planetary aesthetics and 78–89; post- 144–63, 191–2; of post-history 7–17; public (British) 117–30; reciprocal 18–33; urban 104–16; urban models and 65–77
landscape urbanism 7, 73, 177, 185
Landschaftspark Duisburg Nord (Latz) 2
Last Futures (Murphy) 180
Latour, Bruno 19
Latz, Peter 155
Laugier, Marc-Antoine 74, 76 Lawrence Halprin Conservancy 111
Le Corbusier 178
Lefebvre, Henri 2, 156, 157, 189–90
legal approaches 52–64
Le Nôtre, André 11
Life Spent Changing Places, A (Halprin) 105
Linebaugh, Peter 125
Lisbon 131–43, 132*fig*
Local Code (Sorkin) 157
logging 27–8, 57–8
London 144–63
Los Angeles: The Architecture of the Four Ecologies (Banham) 182
Lynch, Kevin 108

'magical extraction' 188–9
mahogany 27
Manifesto for Maintenance Art (Ukeles) 157, 158*fig*
manners 117–30
Marks, Robert 84–5
Massey, Doreen 149, 153, 156
material cultures 8–9
material portraits 18–33
Matta-Clark, Gordon 161, 161*fig*
May, Ernst 178–9
Maya Indigenous Community of Toledo v. Belize 58
McHale, John 179, 180–1, 182
McHarg, Ian 108, 179, 181, 182
McLuhan, Marshall 183
'meanderthal' 122
Mels, Tom 189
Mercator projection 84
Metropolitan Nature Reserve v. Panama 61–2
'microrayon' 95–6
Mies van der Rohe, Ludwig 66, 76
Migge, Leberecht 179
Mitchell, Don 149–50, 156, 157
Mitchell, W.J.T. 189

Mitlin, Diana 169
modernism 179
Modest Witness (Haraway) 86
Moiwana Community v. Suriname 58–9
Morgan, J.P. 23
Mosbach, Catherine 65, 67, 69, 75–6
Moses, Robert 24–5
'Motation' System 104, 106–8, 113–14
Murphy, Douglas 180

National Housing Authority (Thailand) 168
National Industrial Recovery Act (1933) 24
nature, as legal person 63
Nature Conservancy 46
neoliberalism 177–87
New Deal legislation 24, 30
New Materialism 8
New Orleans 34, 35, 36–40, 38*ph*
New York City: Central Park 19–22, 30; High Line 24, 26–30, 28*ph*, 29*ph*, 154; Riverside Park 22–6, 25*ph*, 30; Superstorm Sandy and 34; Union Square 190–1, 192; West Side Improvement 24–6, 30
Non-Sites (Smithson) 19
Novouralsk *see* Sverdlovsk-44/Novouralsk
Now (Akerman) 16*ph*
Nsibidi 110

Occupy London Stock Exchange (LSX) 144, 146*ph*, 147–50, 148*ph*, 152–3, 154, 155, 157, 162, 192
Odd Lots, Revisiting Gordon Matta-Clark's Fake Estates (exhibition) 162, 161*fig*
oikos 78, 80, 82, 86
Olivais 133
Olmsted, Frederick Law 25, 30, 48
Olwig, Kenneth 1, 145, 152
OMA 72–5
open city, concept of 91
'Open City, The' (Sennett) 100
Operating Manual for Spaceship Earth (Fuller) 180
Oppenheimer, Frank 40
Orff, Kate 8, 14
Organische Stadtbaukunst (Organic Urban Design; Reichow) 179
Oxford Circus 117

Para, Brazil 26–30
Parc de la Villette 2, 66
Paris 66, 72, 74–5
Paris COP21 debates 78–9 park projects 65–77
Parque Agricola de Chelas 136*fig*

Parque de Bela Viste 136, 140
Parque Hortícola do Vale do Chelas 135, 137–8, 139–40
Parques Hortícolas 137
Paternoster Square (Whitfield/Farrell) 144, 146*ph*, 147–53, 148*ph*, 152, 155
Penobscot Bay, Maine 19–20
Petrarch 10
physiocracy 11
Picon, Antoine 11
Pittsburgh's Point 22
Placing the Golden Spike: Landscapes of the Anthropocene (exhibition) 9
Plano Urbanização de Chelas (Chelas Urbanization Plan) 131, 132, 133–5, 134*fig*, 136
Plano Verde de Lisboa (Lisbon Green Plan) 137–40
playing card map (Wolff) 44–5, 44*ph*, 45*ph*, 46*ph*
'pocket urbanism' 167, 168
Pollak, Linda 3
Portland Design Week 111
Portland Sequence (Halprin) 104, 105, 110–12, 110*ph*, 111*ph*, 113*ph*
'post-history' 13
privately owned public space (POPS) 125, 148–9
property rights 54–5
propriety 118–19, 122, 124
public space 189–91

Quesnay, Francois 12
queuing 121–2
quintas 136–7, 140

Rahm, Philippe 65, 67, 75–6
Rainforests of New York campaign 29
Rancière, Jacques 165, 172
Rawls, John 165, 172
'reading nature' 12
Recovering Landscape (Corner) 154
Reichow, Hans-Bernhard 179
Remaking London (Campkin) 155
resilience 49, 172, 173–4
Richards, Peter 40
rights: human 52–64; of indigenous people 56–9, 62; to landscape 53–60; of landscape 60–2
River Can 119–20
Riverside Park, New York City 22–6, 25*ph*, 30
Robinson projection 84
Robles, Juan 38*ph*

Roma community 55
Rosati, Clayton 189
Rosin, Scott 38*ph*
Rothstein, Arthur 23*ph*
RSVP cycles (Halprin) 104, 105, 108–10, 114

Salazar, António de Oliveira 131
Sands Quarry 20*ph*
San Francisco Bay 35, 40–2
Sayer, Andrew 121
Scenes from American Deserta (Banham) 182
Schall, Christian 61
Schulze, Mark 27
Schwartzenberg, Susan 40
Seidlungen 178–9
Sennett, Richard 91, 100–1, 123–4
September 11 190–1
Seventeen Contradictions and the End of Capitalism (Harvey) 146–7
Shaler, Nathanial 19–20
shame 121
Shelley, Elise 37, 38
Siddiqi, Asif 102
sidewalks 120–2
'sidewalk zamboni' 122
Sierra Club 108, 110
Skyline Park 108
slavery 80, 86, 87
slavery by debt 27
Sloterdijk, Peter 13
Smithson, Robert 19
Social Formation and Symbolic Landscape (Cosgrove) 1
social justice 31, 165, 170
social order 120–2, 123
Soja, Edward 165, 172
Sorkin, Michael 157
South Gate House 21*ph*
space: abstract 190; differentiated 190; privately owned public (POPS) 125, 148–9; public 189–91
Spang, Chalfant factory 24, 24*ph*
spatial justice 172
Spencer, Douglas 3, 4
Spruce Head Quarry 21
stadtlandschaft 179
Staeheli, Lynn 149
Stalin districts 92, 93
standing (in legal cases) 61, 63
state, development of 11
Steel Workers Organizing Committee 24
Stock Market Crash (1929) 23
Stop Thief! (Linebaugh) 125

Strecker, Amy 4
Street-Porter, Tim 182
streets 117–30
Superstorm Sandy 34
Sustainable Sites Initiative 18
Sverdlovsk-44/Novouralsk 92–6, 94–5*ph*, 101–2

Tabebuia impetiginosa 26–30
Tabebuia serratifolia 26–30
Tagg, John 153, 154
Taichung 65, 69
Taking Part: A Workshop Approach to Collective Creativity 109
Taut, Bruno 178–9
Telles, G. Ribeiro 139
'Terra Fluxus' (Corner) 7, 8
terra nullius doctrine 57
territory 10–11
Thiel, Phillip 108
Tobin Law (1894) 22
Together (Sennett) 123–4
Toronto 18–33, 34
Townscape approach 108
Tree City (OMA) 73–4, 75
Tuolumne River 110

Ukeles, Mierle Laderman 157, 158*fig*, 162
UN Committee on Economic, Social and Cultural Rights 57
UN Convention on International Trade in Endangered Species (CITES) 27
Ungers, O.M. 96
Union Square, New York City 190–1, 192
United States Steel Corporation 23, 24
urban design 65–6, 90–103, 108, 117–30

urban water landscape 39*ph*; see also *Gutter to Gulf*
Utopie 182

'Values, Process and Form' (McHarg) 181
Vauban, Sébastien Le Prestre de 11
Vaux, Calvert 25, 30
Venn, Edwin 151
Vinalhaven 20*ph*, 22
vocabulary/language 34–6, 40–5, 42*ph*, 47, 49

Waldheim, Charles 177
Walking the Elephant 160*fig*
Wall, Ed 161, 191
Walzer, Michael 123, 124
Warren, Waldo 84
Ways of Seeing (Berger) 152
Weizman, Eyal 15–16
West Side Elevated Highway 24–5, 28
West Side Improvement 24–6, 30
Wheatfield: A Confrontation (Denes) 82
Whitfield, William 149, 151
Whole Earth Catalog (Brand) 81, 182
Williams, Raymond 144, 152, 188, 189
'World Energy Map, The' (Fuller) 85
World of Matter 15–16, 15*ph*

Xákmok Kásek Indigenous community 59

Yanomami v. Brazil 58

ZATOs 90–103; *see also* closed cities
Zheleznogorsk *see* Krasnoyarsk-26/Zheleznogorsk
zoe 70, 79